Outstanding Dissertations in
ECONOMICS

A Continuing Garland Research Series

Sulfur Emissions Policies, Oil Prices, and the Appalachian Coal Industry

Robin C. Landis

Garland Publishing, Inc.
New York & London, 1984

Library of Congress Cataloging in Publication Data

Landis, Robin C.
    Sulfur emissions policies, oil prices, and the Appalachian
coal industry.

    (Outstanding dissertations in economics)
    Originally presented as the author's thesis (Ph.D.—
Massachusetts Institute of Technology, 1979)
        1. Coal trade—Government policy—Appalachian Region.
2. Air quality—Government policy—United States.
3. Sulphur oxides—Environmental aspects—Government
policy—United States.   4. Coal-fired power plants—
Government policy—United States.   5. Petroleum
products—Prices—United States.   6. Energy policy—
United States.   I. Title.   II. Series.
HD9546.L35   1984   338.2'724'0974   80-8625
ISBN 0-8240-4182-8

Printed in the United States of America

SULFUR EMISSIONS POLICIES, OIL PRICES,

AND THE APPALACHIAN COAL INDUSTRY

by

Robin Clive Landis

B.A., Political Science and Economics, Yale University
(1966)

SUBMITTED IN PARTIAL FULFILLMENT
OF THE REQUIREMENTS FOR THE
DEGREE OF

DOCTOR OF PHILOSOPHY

at the

MASSACHUSETTS INSTITUTE OF TECHNOLOGY

(April 1979)

Signature of Author...
                          Department of Economics, April 15, 1979

Certified by...
                                              Thesis Supervisor

Accepted by..
                          Chairman, Department of Economics Committee

TABLE OF CONTENTS

TABLE OF CONTENTS (Continued)

Page

TABLE OF CONTENTS (Continued)

TABLE OF CONTENTS (Continued)

Page

LIST OF TABLES

LIST OF TABLES (Continued)

LIST OF TABLES (Continued)

LIST OF TABLES (Continued)

LIST OF TABLES (Continued)

LIST OF TABLES (Continued)

LIST OF TABLES (Continued)

LIST OF FIGURES

Chapter 1

INTRODUCTION AND SUMMARY

This study analyzes the impact of the 1973-1974 oil price increases on the Appalachian coal industry, which otherwise would have suffered large output reductions as a result of sulfur emission restrictions. The study first estimates what 1980 Appalachian coal output and employment would have been under different sulfur emission standards for electric utilities, at relative coal and oil prices before 1973. It then compares these estimates with estimates of 1980 coal output and employment at post-1974 relative prices.

Introduction

The period from the late 1940's to the late 1950's was a period of severe adjustment for the Appalachian coal industry and the regional economy. A sharp decline in coal consumption occurred as natural gas and heating oil displaced coal for home furnaces and as the railroads switched from coal-fired steam engines to diesels. Electric utilities were the only sector in which coal use grew steadily, and by the late 1960's they consumed about half of the coal produced in Appalachia.

1

At the end of the 1960's, however, concern about occupational safety, air quality, and other environmental policies led to legislation causing a substantial increase in the relative cost of burning coal. The Federal Coal Mine Health and Safety Act went into effect in 1969, with the purpose of bringing about long-overdue reforms in underground coal mining practices, but with the necessary result of increasing mining costs. A variety of federal and state legislation deals with acid mine drainage from coal mines.[1] Congress has been considering legislation to impose national reclamation standards for strip mines.

As of this time, however, it is the air quality standards that seem likely to bring about the largest increase in the relative cost of burning coal. The 1967 Air Quality Act, as amended in 1970, provides for a complex set of standards to be met or exceeded by state regulations.[2] The EPA guidelines for sulfur emissions for electric utilities roughly imply a 0.7 percent sulfur-in-fuel standard. Technology currently exists for desulfurizing residual fuel oil to this level, but stack gas treatment will probably be required to burn coal in compliance with this standard.

---

[1]Policies dealing with health and safety, water quality, and reclamation of strip-mined lands are discussed in Charles River Associates, Incorporated, The Impact of Public Policy on the Appalachian Coal Industry and the Regional Economy, Volume II, "Impact of Environmental and Other Policies on the Appalachian Coal Industry", pp. 101-195.

[2]The complex of regulations is described in detail in Charles River Associates, op. cit., pp. 7-29.

With the known but not-yet commercially proven technology, stack gas treatment appears to be expensive.[1]

At relative oil and coal prices prior to the embargo and price increases of 1973-1974, imposition of a strict sulfur-in-fuel standard would have increased sharply the costs of burning coal. In this study, I estimate the impact on Appalachian coal output and employment of such a cost increase, and I compare its costs and impacts with those of a less stringent sulfur-in-fuel standard.

I then estimate 1980 Appalachian coal output and employment under the assumption that the 1973-1974 oil price increases are maintained, considering both the resulting slow-down in electricity consumption and the improved competitive position of coal. Broadly, I conclude that the oil price increases engineered by the Organization of Petroleum Exporting Countries (OPEC) rescued the Appalachian coal industry from output reductions attributable to sulfur emissions standards.

---

[1]The status of alternate control processes is discussed in Charles River Associates, op. cit., pp. 80-96. More recent cost estimates are summarized in Office of Fuel Utilization, Federal Energy Administration, Final Environmental Statement, Coal Conversion Program (April 1975), pp. IV-110--IV-118.

3

Summary

The rest of this study is divided into six chapters. In Chapter
2, I survey the demand for Appalachian coal, tracing out post-war
consumption patterns. Chapter 3 is devoted to a closer look at the
competition between Appalachian coal and other fossil fuels -- primarily
Western and Midwestern coal, natural gas, and imported residual fuel
oil.

In Chapter 4, I develop a model of coal use by electric utilities.
This model is used in Chapter 6 to evaluate the impact of alternative
sulfur-in-fuel standards. In Chapter 5, I analyze and forecast coal
consumption in end-uses other than electric utilities. Consumption is
considered separately for coking, exports, retail demand, and other
manufacturing and mining. In Chapter 7, I forecast Appalachian coal
output and employment at post-embargo relative prices, focusing on
conversions from oil to coal and the attendant costs.

There are two main conclusions. First, the costs of compliance with
a 2.0 percent sulfur-in-fuel standard are roughly one-third those of a
0.7 percent standard. A 2.0 percent standard is also much more
favorable to output and employment in the Appalachian coal industry.
The additional costs of a 0.7 percent standard are great enough that the
benefits of the stricter standard should be carefully evaluated.

4

Second, the oil price increases of 1973-1974 effectively rescued the Appalachian coal industry from the strict sulfur-in-fuel standard. The output decline, in the absence of the oil price increases, would not have been as cataclysmic as that following World War II, but it would have been substantial. Even with the slow-down in electricity consumption resulting from higher electric rates, the improved competitive position of coal relative to oil implies a level of coal consumption roughly equal to that which would have occurred without the sulfur emission restrictions.

Chapter 2

SURVEY OF DEMAND FOR APPALACHIAN COAL

At present there are only a few major uses for coal. The largest
single use, as fuel for electric utilities, is also the only domestic
one with prospects for substantial growth. The other main uses -- as
the basic input for coke for the steel industry and as fuel for power and
process steam in a variety of industries -- have grown little if at all
during the postwar period. However, they have not been threatened
directly by competitive fuels.[1]

The utility market, on the other hand, is intensely competitive
in its choice of fuels, and the current patterns of fuel use result
directly from the relative delivered cost per BTU of the various
alternatives -- coal, natural gas, and residual fuel oil. These costs,
in turn, depend on production costs, transporation costs, and government
regulations.

---

[1]The demand for coal in these uses is discussed in greater detail
in Chapter 4.

Because transportation costs per BTU are so high for coal, distance from the coal fields is an important determinant of delivered price. For this reason, coal from the Far West (Colorado, New Mexico, Wyoming, Montana) does not compete directly with Appalachian coal.[1] Interfuel competition (for Appalachian coal) occurs only on three geographical margins -- along the eastern seaboard, where it competes with imported residual fuel oil; in the east north central section, where it competes with midwestern coal; and in the east south central region, where it competes with natural gas.

In this chapter, I survey briefly the demand for Appalachian coal. I consider, in turn, current patterns of coal use, historical trends in coal consumption, and competitive fuel positions and prospects.

Current Patterns of Coal Consumption

Coal is used as an energy source for the production of steam and as the raw material for coke. In addition to its consumption by electric power and coke industries, coal is consumed by steel and cement mills, by home owners, and by a heterogeneous group called "other manufacturing and mining industries".

---

[1]As discussed later in this chapter, since 1972 some Western coal has been burned in the traditional markets of Appalachian coal. Thus far, this substitution has been largely at the expense of Midwestern rather than Appalachian coal; if stack gas desulfurization proves ineffective or too costly, however, Western coal may displace high sulfur Appalachian coal.

The chemical composition of coal varies considerably, not only across ranks (such as bituminous, sub-bituminous, anthracite, and lignite) but also within each rank. The characteristics of coal that determine its market price include heat content, ash content, sulfur content, and ability to "coke". The ash fusion temperature determines whether the coal can be burned in a wet-bottom boiler, but does not directly affect the price of the coal. (The relationship between boiler type and coal characteristics is discussed more fully in Chapter 4.)

Prior to sulfur emission restrictions, the BTU content of the coal was by far the most important consideration for the boiler fuel market, which essentially purchases heat. Ash content and sulfur content do affect the efficiency with which coal can be burned, but, historically, they have been of secondary concern. Coal for coking, on the other hand, should be low in sulfur (to transmit minimal impurity to the iron) and low in ash, and low volatile (in addition, of course, to being *able* to coke (midwestern coals, for instance, coke much less readily than Appalachian coals). Considerable blending of coals is usually done to obtain an optimal composition for coking, however, so that there is not a unique "coking coal".

In the past, the demand for coal for coking did not directly affect the demand for coal for boiler fuel, because of the different properties required. The introduction of sulfur regulations, however, to the extent that they are effectively enforced, puts both utilities and coke producers in the same market for low sulfur coal, causing the price of such coal to rise.

Table 2-1 shows the 1975 pattern of shipments, by end use, of Appalachian and non-Appalachian coal. Even though coal for coke and gas plants is a single category, the dominance of Appalachian coals for coking purposes stands out clearly. Virtually all coal for overseas export comes from Appalachia (primarily from southern West Virginia) because of low transport costs to seaports and the coal's relatively high value-to-weight ratio. Slightly less than one-half of the coal burned by electric utilities came from Appalachia, with West Kentucky and the Midwest shipping most of the rest.

In 1975, electric utilities received almost 70 percent of continental coal shipments. Coke and gas plants accounted for 14 percent, while "other" domestic uses took less than 10 percent. The electric utility market, then, is by far the most important consumer of coal. As will be discussed below, this market has grown fastest and, in spite of the post-1973 slow-down in electricity consumption, is expected to grow fastest, at least to 1980.

The consumption technology of alternative fuels concerns the future demand for coal mainly in the electric utility market, the area of active interfuel competition. The technology of burning natural gas is the simplest. Gas requires no handling or preheating but is burned directly. It is almost completely oxidized in the combustion process, so that there is no residue to be disposed of.

9

Table 2-1

CONTINENTAL SHIPMENTS OF COAL BY END USE, 1975[1]
(Millions of Tons)

	Electric Utilities	Coke and Gas	Retail	Overseas Exports[2]	Others	Total
Appalachian	218.3	81.9	3.5	48.2	36.4	388.2
Non-Appa-lachian	220.3	10.6	1.5	0.2	19.9	252.6
TOTAL	438.6	92.5	5.0	48.4	56.3	640.8

Regional Share in Each Category, Percent

Appalachian	49.8	88.5	69.4	99.5	64.6	60.6
Non-Appa-lachian	50.2	11.5	30.6	0.5	35.4	39.4

[1]Includes shipments to the United States from Canada and Mexico and shipments from United States to Canada and Mexico.

[2]Excludes Canada.

SOURCE: U.S. Bureau of Mines, *Mineral Industry Surveys*, "Bituminous Coal and Lignite Distribution, Calendar Year 1975", April 12, 1976, pp. 4-5.

10

Residual fuel oil, likewise, leaves no ash, but storage tanks are needed, as is equipment to preheat the heavy oil so that it will burn efficiently. Burning coal, on the other hand, requires the most extensive ancillary equipment, including access for rail cars, storage space, coal handling and crushing equipment, and ash disposal equipment.[1]

The implication of this hierarchy of complexity is that it is relatively inexpensive to convert a coal burning furnace to an oil or gas burning furnace, or an oil burning furnace to a gas burning one, and it will almost always pay to do so if the fuel cost per BTU of oil or gas is less than that of coal and is expected to remain so. Conversion in the opposite direction, however, is more costly, and the price per BTU of coal must be considerably below that of oil or gas to justify the installation of the required equipment.

Reconversion costs, estimated below, depend on whether the plant has retained the coal handling facilities or not. In some cases, the plant may have been expanded after the removal of the coal equipment so that, on the available site, reconversion would require a reduction in generating capacity. Where the equipment has been maintained, reconversion would not be expensive.

---

[1]A more complete listing of the additional facilities might include: storage yards, car dumpers, crusher houses, conveyor systems, pulverizers, scales, combusion air systems, ash handling facilities, disposal facilities, and fly-ash collectors. See, for example, Thomas Browne, "Interruptions of Residual Fuel Oil Supplies for Electric Generation -- A Potential Threat to the State's Electric Power Supplies", Office of Economic Research Report No. V (New York State Public Service Commission, November 11, 1971), p. 8.

11

Coal conversion costs are specific to individual plants, depending on the extent and state of repair of coal handling and burning equipment, space available for needed construction, availability of coal transportation facilities, and so forth. A recent attempt to quantify these costs in general assigned capital costs of $7.50 per kilowatt for restoration of coal handling facilities, $20.00 per kilowatt to upgrade electrostatic precipitators, and additional operating costs for coal of 0.2 mills per kilowatt-hour.[1] Translation of capital charges into equivalent fuel costs depends, in turn, on the remaining plant life, interest rates, heat rates (BTU's per kilowatt hour), and plant load factors (capacity utilization rates). For the plants listed by the Federal Energy Administration as able to convert, I estimated that these conversion costs were equivalent to a fuel

---

[1]Office of Fuel Utilization, Federal Energy Administration, Final Environmental Statement, Coal Conversion Program, Energy Supply and Environmental Coordination Act of 1974, Section 2 (April 1975), pp. IV-109--IV-118.

These estimates are substantially below estimates of conversion costs made for three New England plants by the Center for Energy Policy, Inc., The Impact of Power Plant Coal Conversion on New England Energy Policy (Boston, MA: May 1976). Their estimates ranged from $30.75 per kilowatt for Somerset to $51.20 per kilowatt for Brayton Point for conversion alone, with additional electrostatic precipitator upgrading costs ranging from $18.37 per kilowatt for Mt. Tom to $90.00 per kilowatt for Somerset. Their estimated operating and maintenance charges ranged from nil for Somerset to 1.36 mills per kilowatt-hour for Brayton Point. There appears to be a very wide variance in these estimates, and with so few plants it is difficult to evaluate the representativeness or the validity of the estimates. These figures do suggest that there is a great deal of uncertainty about conversion costs.

charge ranging from 7.2 to 23.5 cents per million BTU.[1] These costs,
then, are not negligible even for plants that had the capacity to burn
coal as of June 22, 1974.

Postwar Trends in Coal Consumption

The structure of coal demand has changed dramatically since the
end of World War II. Two sets of events have exerted opposite influences
on the level of coal consumption, with the net result that, by 1969,
coal consumption was still below its 1947 level and the composition of
demand had shifted markedly.

Table 2-2, showing coal consumption by selected consumer classes,
reveals the changes in composition. Two categories -- Class I railroad
and retail consumption -- declined, while consumption by electric
utilities grew. The displacement of coal from the home heating market
by distillate fuel oil and by natural gas was due largely to the
superior cleanliness and convenience of the latter two fuels, coupled
with the development of a pipeline network for natural gas. The
adoption of diesel locomotives led to a virtual disappearance of consumption by this once-dominant source of coal demand.

By 1957, a year of general prosperity, coal consumption was about
400 million tons. The economic stagnation of the next four years
reduced annual consumption to about 375 million tons. After 1961, however,
coal consumption steadily increased as a result of electric utility use.

---

[1]The methods are discussed in detail in Chapter 7.

Table 2-2

YEARLY CONSUMPTION, BY CONSUMER CLASS,

OF BITUMINOUS COAL, 1935-1975
(Millions of Tons)

Year	Electric Power Utilities	Bunker Foreign Trade	Railroads (Class 1)	Beehive Coke Plants	Oven Coke Plants
1935	30.9	2.7	77.1	1.5	49.0
1940	49.1	3.0	85.1	4.8	76.6
1945	71.6	3.1	125.1	8.1	87.2
1950	88.3	2.0	61.0	9.1	94.8
1955	140.6	1.5	15.4	2.9	104.5
1956	155.0	1.5	12.3	4.0	101.9
1957	157.4	1.3	8.4	3.5	104.5
1958	152.9	1.0	3.7	1.0	75.6
1959	165.8	1.0	2.6	1.8	77.3
1960	173.9	0.9	2.1	1.6	79.4
1961	179.6	0.8	[1]	1.5	72.4
1962	190.8	0.7	[1]	1.3	72.9
1963	209.0	0.7	[1]	1.6	76.0
1964	223.0	0.7	[1]	2.0	86.7
1965	242.7	0.7	[1]	2.7	92.1
1966	264.2	0.6	[1]	2.4	93.5
1967	271.8	0.5	[1]	1.4	90.9
1968	294.7	0.4	[1]	1.3	89.5
1969	308.5	0.3	[1]	1.2	91.7
1970	320.5	0.3	[1]	1.4	94.6
1971	326.3	0.2	[1]	1.3	81.5
1972	348.6	0.2	[1]	1.1	86.2
1973	386.9	0.1	[1]	1.3	92.3
1974[2]	390.1	0.1	[1]	1.3	88.4
1975[2]	403.2	-	[1]	1.1	82.2

[1]Canvass discontinued

[2]Preliminary

Table 2-2 (Continued)
YEARLY CONSUMPTION, BY CONSUMER CLASS, OF BITUMINOUS COAL, 1935-1975
(Millions of Tons)

Year	Steel and Rolling Mills	Cement Mills	Other Manufacturing and Mining Industries	Retail Dealer Deliveries	Total
1935	16.6	3.5	94.6	80.4	356.3
1940	14.2	5.6	107.9	84.1	430.1
1945	14.2	4.2	126.6	119.3	559.6
1950	10.9	7.9	95.9	84.4	454.2
1955	7.4	8.5	89.6	53.0	423.4
1956	7.2	9.0	93.3	48.7	432.9
1957	6.9	8.6	87.2	35.7	413.9
1958	7.3	8.3	81.4	35.6	366.7
1959	6.7	8.5	73.4	29.1	366.3
1960	7.4	8.2	76.5	30.4	380.4
1961	7.5	7.6	77.3	27.7	374.4
1962	7.3	7.7	78.8	28.2	387.8
1963	7.4	8.1	82.8	23.5	409.2
1964	7.4	8.7	82.9	19.6	431.1
1965	7.5	8.9	85.6	19.0	459.2
1966	7.1	9.1	89.3	20.0	486.3
1967	6.3	8.9	83.5	17.1	480.4
1968	5.7	9.4	82.6	15.2	498.3
1969	5.6	[1]	85.4[2]	14.7	507.3
1970	5.4	[1]	82.9	12.1	517.2
1971	5.6	[1]	68.7	11.4	494.9
1972	4.9	[1]	67.1	8.7	516.8
1973	6.4	[1]	60.8	8.2	556.0
1974[3]	6.2	[1]	57.8	8.8	552.7
1975[3]	2.7	[1]	59.8	5.7	554.7

Footnotes and Sources on next page of this table.

15

Table 2-2 (Continued)

YEARLY CONSUMPTION, BY CONSUMER CLASS,
OF BITUMINOUS COAL, 1935-1975
(Millions of Tons)

[1]From 1969 on, consumption by cement mills is included in "Other Manufacturing and Mining Industries".

[2]From 1969 on, includes consumption by cement mills.

[3]Preliminary

SOURCES: 1935-1968: National Coal Association, *Bituminous Coal Facts/1969 edition*, p. 82.

          1969-1971: *Ibid.*, 1971 edition, p. 77.

          1972-1973: U.S. Bureau of Mines, *Mineral Industry Surveys*, "Coal-- Bituminous and Lignite in 1973" (January 4, 1975), p. 62.

          1974-1975: U.S. Bureau of Mines, *Weekly Coal Report*, April 9, 1976, p. 8.

Determination of coal's precise share of the electric utility market is complicated by the slightly different coverages of *Steam-Electric Plant Factors* and FPC compilations, as well as by the choice of a BTU content to convert each fuel's consumption into comparable units. Further, slight amounts of anthracite are also burned, and the two sources may treat these amounts differently. In general, however, all of the series move together, so that trends are unambiguous. Table 2-3 shows consumption in physical units of coal, residual fuel oil, and natural gas by electric utilities, and Table 2-4 shows the conversion of these quantities into heat contents and each fuel's share of the total.[1] Table 2-4 reveals that between 1950 and 1965 coal not only dominated the electric utility fuel market but also maintained its share, even in the face of a more rapid increase in natural gas consumption.

Electric utilities in areas close to Appalachia mainly consume gas on an interruptible basis, although until recent years interruptions were not very frequent.[2] The reduced price for interruptible gas, coupled with FPC regulation of the field price of gas, makes gas, where available, the cheapest source of BTU's. The growth of the pipeline network made possible the distribution of production from the vast reserves discovered in connection with oil exploration and development.

---

[1]The conversion factor for coal overstates the BTU content of this fuel in recent years, but an adjustment would not change the main conclusions.

[2]"Interruptible", as used here, refers to gas delivered at the option of the pipeline or gas utility, usually during off-peak periods (such as summer). It is different from "firm" gas, for which both the amount and the time of delivery are specified by contract.

17

Table 2-3

FUEL CONSUMPTION BY ELECTRIC UTILITIES,
FISCAL YEARS, 1947-1974

Year	Coal (Millions of Tons)	Oil (Millions of Barrels)	Gas (Millions of Cubic Feet)
1947	89.5	45.3	373,054
1948	99.6	42.6	478,097
1949	84.0	66.3	550,121
1950	91.9	75.4	628,919
1951	105.8	63.9	763,898
1952	107.1	67.2	910,117
1953	115.9	82.2	1,034,272
1954	118.4	66.7	1,165,498
1955	143.8	75.3	1,153,280
1956	158.3	72.7	1,239,311
1957	160.8	79.7	1,336,141
1958	155.7	77.7	1,372,853
1959	168.4	88.3	1,628,509
1960	176.6	85.3	1,724,762
1961	182.1	85.7	1,825,117
1962	193.2	85.8	1,965,974
1963	211.3	93.3	2,144,473
1964	225.4	101.1	2,322,896
1965	244.8	115.2	2,321,101
1966	266.5	140.9	2,609,949
1967	274.2	161.3	2,746,352
1968	296.9	187.9	3,143,858
1968[1]	292.8	171.9	3,047,990
1969	306.4	232.6	3,358,340
1970	319.6	302.9	3,719,443
1971	327.4	350.7	3,832,550
1972	344.5	429.7	3,714,731
1973	384.6	496.6	3,361,185
1974	393.6	476.4	3,178,738

Footnotes and Sources on next page of this table.

18

Table 2-3 (Continued)

FUEL CONSUMPTION BY ELECTRIC UTILITIES,
FISCAL YEARS, 1947-1974

[1]Data from 1968 to 1973 are on a calendar rather than fiscal
year basis.

SOURCES: 1947-1968:   Federal Power Commission, *Annual Reports*,
                        collected in American Petroleum Institute,
                        Petroleum Facts and Figures (1971), p. 440.

           1968-1974:   National Coal Association, *Steam-Electric
                        Plant Factors*, 1969 to 1975 editions, Table 2.

19

Table 2-4

HEAT EQUIVALENTS OF FUELS CONSUMED IN THE GENERATION OF POWER
BY ELECTRIC UTILITIES, FISCAL YEARS 1947-1974

Year	BTU's			Percent		
	Coal	Oil	Gas	Coal	Oil	Gas
1947	2,148.7	284.9	385.0	76.2	10.1	13.7
1948	2,390.1	268.1	493.3	75.8	8.5	15.7
1949	2,015.1	416.8	567.7	67.2	13.9	18.9
1950	2,204.9	474.2	649.0	66.3	14.2	19.5
1951	2,538.4	402.0	788.3	68.1	10.8	21.1
1952	2,569.7	422.6	939.2	65.4	10.7	23.9
1953	2,781.5	517.0	1,067.4	63.7	11.8	24.4
1954	2,841.2	419.6	1,202.8	63.7	9.4	26.9
1955	3,450.2	473.2	1,190.1	67.5	9.3	23.3
1956	3,798.7	457.1	1,279.0	68.6	8.3	23.1
1957	3,858.5	501.0	1,379.0	67.2	8.8	24.0
1958	3,737.4	488.3	1,416.8	66.2	8.7	25.1
1959	4,042.1	554.9	1,680.6	64.4	8.8	26.8
1960	4,239.2	536.5	1,780.0	64.7	8.2	27.2
1961	4,370.9	539.0	1,883.5	64.3	7.9	27.7
1962	4,637.7	539.2	2,028.9	64.4	7.5	28.2
1963	5,072.0	586.7	2,213.1	64.4	7.4	28.1
1964	5,410.2	635.9	2,397.2	64.1	7.5	28.4
1965	5,874.9	724.2	2,395.4	65.3	8.1	26.6
1966	6,395.5	886.1	2,693.5	64.1	8.9	27.0
1967	6,580.4	1,014.0	2,834.2	63.1	9.7	27.2
1968	7,126.6	1,181.4	3,244.5	61.7	10.2	28.1
1968[1]	6,890.7	1,075.6	3,177.3	61.8	9.7	28.5
1969	7,133.2	1,450.6	3,484.5	59.1	12.0	28.9
1970	7,229.2	1,880.3	3,849.6	55.8	14.5	29.7
1971	7,312.1	2,166.1	3,958.6	54.4	16.1	29.5
1972	7,705.0	2,640.3	3,834.2	54.3	18.6	27.0
1973	8,529.3	3,041.2	3,457.1	56.8	20.2	23.0
1974	8,557.4	2,918.0	3,267.1	58.1	19.8	22.2

Footnotes on following page.

Table 2-4 (Continued)

HEAT EQUIVALENTS OF FUELS CONSUMED IN THE GENERATION OF POWER
BY ELECTRIC UTILITIES, FISCAL YEARS 1947-1974

[1]1968 to 1974 are calendar year figures.

UNITS: Coal, oil and gas BTU's in trillions

           Coal, oil and gas percentages as percentages of total shown.

SOURCE: 1947-1968 computed from Table 2-3, according to the following
           BTU equivalents:
           -- 1 short ton of coal yields 24,000,000 BTU
           -- 1 barrel of residual yields 6,287,000 BTU
           -- 1 cubic foot of natural gas yields 1,032 BTU

           1968-1974: *Steam-Electric Plant Factors* (full citation in
           Table 2-3)

21

The decline in coal's share after 1966 is due almost exclusively
to the effective exemption, in that year, of residual fuel oil from
the oil import quota for Petroleum Administration for Defense (PAD)
District 1.[1] The low price of imported residual oil, pollution
regulations, and the ease of shifting from coal-burning to oil-burning
led to a rapid shift away from coal by utilities close to or on the
seacoast. Coal's share in New England fell from 63 percent in 1964
to 11 percent in 1974; in the Middle Atlantic states it fell from 77
percent to 56 percent, and in the South Atlantic states from 79 to
63 percent over the same period. In all other regions coal's share
changed little. Another indication of this shift was the increase
in oil consumption from 115 million barrels in 1965 to 188 million
barrels in 1968 and to 476 million barrels in 1974.

The shift has been most visible in the large cities on the East
Coast, with their large ports and stringent sulfur emission restrictions.
For example, in 1974 utilities in New York City burned 94 percent oil and
no coal, but utilities in the rest of the state burned 52 percent coal
and 46 percent oil. In Philadelphia, utilities burned 63 percent oil
and 35 percent coal, but utilities in the rest of the state burned
98 percent coal. In spite of the quadrupling of world oil prices
after 1973-1974, no plants in large East Coast cities have reconverted
to coal (and some coal-burning plants have converted to oil). The

---

[1]PAD District 1 includes Maine, Vermont, New Hampshire, Massachusetts,
Connecticut, Rhode Island, New York, Pennsylvania, New Jersey, Maryland,
Delaware, West Virginia, North Carolina, South Carolina, Georgia, and
Florida.

only large East Coast cities in which some plants have been ordered by the FEA to reconvert to coal are Baltimore, Maryland and Wilmington, Delaware. As of 1974, no new coal-burning plants are listed as planned for East Coast cities.

This discussion of the demand for fuel by electric utilities has focused, so far, only on conventional steam-electric plants. Two other methods of generating electricity, competitive with conventional thermal plants, need to be mentioned -- hydroelectric and nuclear. Hydroelectric power is important in only a few regions, and the consensus is that there are few, if any, economical sites left for the construction of additional hydro capacity.[1] For this reason no further consideration is given it here.

Nuclear generation, with its low operating costs and negligible fuel transport costs, has long been heralded as the eventual primary source of electricity. As discussed in Chapter 3, the threat of displacement from the utility market by nuclear power has slowed down the expansion of coal capacity and has contributed to the use of long-term contracts between utilities and coal producers.

Although a commercial nuclear power plant was in operation as early as 1957, the growth in nuclear power has been much slower than originally expected. Increases in construction costs, delays, labor troubles, and difficulties encountered in expanding the scale of plants are some of the reasons for the slowness.

---

[1]This is particularly true in areas where Appalachian coal is competitive.

Further delays may result from opposition to possible adverse environmental effects from nuclear power plants. The revised Atomic Energy Commission rules governing environmental impacts will have to be taken seriously, in view of the Calvert Cliffs decision, which requires the AEC to weigh the benefits and costs of alternative plant sites and designs. The decision applies, moreover, to all nuclear plants for which an operating license has not yet been granted.[1] Because nuclear plants impart a great deal of heat to the environment, plant designs that raise the temperature of nearby bodies of water by an acceptable amount may require more time or money to implement.

As of 1974, for example, fossil generating capacity was on the order of 339,000 megawatts (MW), while installed nuclear capacity was about 32,000 MW.[2] Even if allowances are made for higher nuclear load factors and conservative rating of nuclear capacity, these figures still imply that nuclear generation is less than 10 percent of conventional generation. Moreover, with the delays in the last few years, the expected lead time of a nuclear plant announced after 1968 is 9 to 10 years from date of announcement to date of commercial operation. Table 2-5 shows, by announcement date, the average months required (or expected) for various stages of the process from announcement to operation. Because of the

---

[1]*Calvert Cliffs Coordinating Committee v. AEC*, 449 F. 2d 1109.

[2]National Coal Association, *Steam-Electric Plant Factors, 1975 Edition* (Washington: January 1976), pp. 44-45, 53.

Table 2-5

AVERAGE LEAD TIMES, NUCLEAR POWER PLANTS,
BY ANNOUNCEMENT DATE (IN MONTHS)

Announce-ment Date	Number of Plants	From Announce-ment to AEC Permit	From AEC Permit To Start of Con-struction	From Start of Con-struction to Commercial Operation	Total, From Announcement To Commercial Operation
1959	2	6.0	11.5	65.5	83.0
1960	--	--	--	--	--
1961	--	--	--	--	--
1962	--	--	--	--	--
1963	3	3.0	11.3	54.7	69.0
1964	--	--	--	--	--
1965[1]	8	3.9	11.4	54.4	69.5
1966	19	3.9	12.8	56.1	72.8
1967	28	8.9	19.3	74.1	102.1
1968	14	10.3	24.7	69.0	104.0
1969[3]	7	7.4	32.4	65.0	119.4
1970[3]	11	9.9	30.2	68.9	105.4
1971[3]	18	11.8	2-.0	64.8	116.2
1972[3]	27	14.0	19.0	65.5	119.4
1973[3]	31	10.2	20.0	85.0	119.9
1974[3]	22	0.8	NA[2]	NA[2]	112.5

NOTES:

[1]From 1965 on, averages include expected lead times as well as actual ones.

[2]"NA" means no estimates were given for these stages.

[3]The averages for entries in these years are based on fewer than the total number of plants announced. The actual number of plants for which estimates were given appear on the next page of this table.

Table 2-5 (Continued)

AVERAGE LEAD TIMES, NUCLEAR POWER PLANTS,
BY ANNOUNCEMENT DATE (IN MONTHS)

Number of Plants on which Average Lead-Times are Based

Announcement Date	Announcement to Permit	Permit to Construction	Construction to Operation
1969	7	5	5
1970	11	9	9
1971	18	6	6
1972	27	2	2
1973	26	1	1
1974	14	–	–

SOURCES: 1959-1966: Compiled from *Electrical World*, October 15, 1971, p. 40, based on Southern Nuclear Engineering, Inc., *Commercial Nuclear Power Plants* Edition No. 4 (Dunedin, FL: October 1971).

1967-1974: Compiled from *Electrical World* (October 15, 1974) pp. 42-50; based on Southern Nuclear Engineering Data

lead times involved, and because of limits on the capacity of nuclear construction equipment and personnel, it is most likely that only plants announced by 1971 will be in operation by 1980. Moreover, further slippages may occur, so that even this figure may overstate the actual expansion of nuclear capacity.

The upper limits on the growth in nuclear capacity to 1980 are, therefore, well established. In Chapter 4, I discuss the assumption made about the timing of nuclear additions.

Low Sulfur Western Coal

Appalachian coal competes with other fuels primarily in the electric utility market and, even there, only on certain geographical margins. Historically, residual fuel oil and natural gas have been coal's main fossil fuel competitors. Sulfur emission restrictions, along with the uncertainty and expense of flue gas desulfurization devices, have enhanced the competitiveness of low sulfur Western coal in many of Appalachian coal's market areas.

As of 1975, shipments of Appalachian coal to utilities had been little affected by competition from low sulfur Western coal. An exception to this general statement is Detroit Edison, which has begun to burn substantial amounts of Western coal. Although that consumption in 1975 was not large enough to lower appreciably the share of Appalachian coal in coal consumption by Michigan electric utilities, by the first nine months of 1976, consumption of Western coal in Michigan accounted for about 12 percent of total utility

27

consumption. As compared with the corresponding period in 1975, the share of Appalachian coal consumption fell from about 87 to 80 percent.[1]

Because some Western coal is currently being burned by utilities in other states that also burn Appalachian coal, in this section I briefly examine competition from Western coal. I do not attempt to determine how much Western coal will eventually be shipped east of the Mississippi, but rather examine the current status and sketch the determinants of future developments.

Large amounts of Western coal are low in sulfur and can be stripmined at a low cost per ton.[2] Thus, in spite of high transport costs to Eastern markets, burning Western coal may prove an economical way to comply with sulfur emission restrictions, especially if flue gas desulfurization proves unreliable or too expensive. American Electric Power, for example, a large electric utility located mainly in Appalachia, has recently signed long term contracts for low sulfur Western coal.[3]

---

[1]Computed from shipments data in Bituminous Coal and Lignite Distribution, January-September 1976 and January-September 1975.

[2]A fuller discussion of the determinants of Western coal price can be found in Martin Zimmerman, Long Run Mineral Supply: The Case of Coal in the United States, unpublished Ph.D. dissertation, M.I.T. (September 1975), pp. 84-89.

[3]Ibid., p. 90.

The Tennessee Valley Authority, citing its inability to obtain low sulfur Appalachian coal, bought 405 thousand tons of southeastern Montana stripmined coal on the spot market. The coal, destined for a plant in Western Kentucky, cost $19 per ton delivered, of which about two-thirds was transportation costs.[1]

Although Western coal output grew by 39 percent between 1971 and 1973 (while total U.S. coal output increased by only 7 percent), total output in that year was less than 10 percent of the contiguous United States total.[2]

Most Western coal is, currently, being burned by utilities outside of Appalachian coal's market areas. For example, in the 24 states in which utilities burned some Appalachian coal in 1969, 22.3 million tons of Western coal were received by utilities in 1975. Of this total, 19.2 million, or about 86 percent was received by utilities in Illinois, Indiana, and Wisconsin, which, even in 1969, received only 0, 1.7 and 7.6 percent of their coal, respectively, from Appalachia.[3]

---

[1]*Keystone News Bulletin,* Vol. 32, No. 12, December 27, 1974, p. 2.

[2]*Ibid.,* Vol. 32, No. 10, October 28, 1974, p. 3. The Western states whose output is included in this comparison are Arizona, Colorado, Montana, New Mexico, North Dakota, Utah, Washington and Wyoming.

[3]Illinois utilities in 1969 received some, but less than 0.5 percent of their coal from Appalachian districts. The 1975 data are from U.S. Bureau of Mines, *Mineral Industry Surveys,* "Bituminous Coal and Lignite Distribution", 1975.

Table 2-6 , which compares the 1970 and 1975 share of Appalachian coal burned by utilities in states accounting for 22.1 of the 22.3 million tons of Western coal, shows that the share of Appalachian coal fell between 1970 and 1975 in only one of the five states.

It appears, therefore, that Western coal has yet to make significant inroads into Appalachian coal markets. In the future, however, it may, depending on a number of factors. It is too early to evaluate these factors, but they are easy to identify. They may be classed in two groups - constraints on Western coal supply (including transportation constraints), and constraints on utilities' ability to burn Appalachian coal.

The major constraint on Western coal development appears, at the present time, to be pending stripmining legislation. For example, requiring operators to restore original vegetation may prove prohibitively costly, while a provision requiring them to restore only the original contour would impose low additional costs.[1] Transportation represents a substantial portion of the delivered cost of Western coal, and if shipments increase substantially the existing rail network could prove a bottleneck, raising transportation costs.

Constraints on the ability of utilities to burn Appalachian coal may arise if enforcement of the Clean Air Act requires either fuel with less than 0.7 percent sulfur or else stack gas desulfurization.[2] Since

---

[1]The discussion in this and the next paragraph relies heavily on a conversation with Martin Zimmerman of the Energy Laboratory at M.I.T.

[2]Of utility plants in compliance with sulfur emissions requirements in 1974, for example, 68 percent of the coal burned contained more than 1.0 percent sulfur, and 37 percent contained more than 2.0 percent (Environmental Protection Agency, *Report to Congress on Control of Sulfur Oxides,* EPA-450/1-1-75-001, February 1975, p. 14.

Table 2-6

SHARE OF SHIPMENTS FROM APPALACHIAN PRODUCING DISTRICTS
TO EAST NORTH CENTRAL UTILITIES, 1970 AND 1975

State	1970	1975	1975 Western Coal Receipts[1]
Illinois	*	0.005	11,651
Indiana	0.017	0.047	3,958
Ohio	0.928	0.930	1,386
Michigan	0.936	0.885	1,502
Wisconsin	0.076	0.096	3,567

*Less than 0.005

[1]Western Coal Receipts in Thousands of Tons

SOURCE: U.S. Bureau of Mines, *Mineral Industry Surveys,* "Bituminous
          Coal and Lignite Distribution", 1970 and 1975.

the supply of low sulfur coal from Appalachia appears quite inelastic, if stack gas desulfurization proves ineffective or very costly relative to the differential in delivered cost per BTU between Western and Appalachian coal, utilities may choose to burn Western coal as the lowest cost fuel in compliance with the sulfur emissions requirement.[1] Another consideration in this regard is that Western coals typically have a lower BTU content than Appalachian coals. Consequently, when burned in existing boilers designed for high BTU coal, the effective generating capacity of the unit decreases.

Although this choice could, in principle, by analyzed with the model developed in Chapter 4, a detailed analysis would require work beyond the scope of the study. Because the uncertainties are very great, in Chapter 7 I present only a very rough analysis based on a comparison of the costs of Western low sulfur coal at a few locations in the market area of Appalachian coal with the costs of Eastern coal plus desulfurization costs.

---

[1]The inelasticity of supply of low sulfur Eastern coals is shown in Zimmerman, op. cit., Chapter VI. This is not, of course, formally inconsistent with the assumption made in this study that the supply of Appalachian coal, taken all in all, is elastic, since low sulfur reserves account for only between an estimated 10 and 20 percent of potential 1980 Appalachian coal output (Environmental Protection Agency, *Report to the Congress on Control of Sulfur Oxides*, p. 20).

Chapter 3

PATTERNS OF COMPETITION IN APPALACHIAN COAL MARKETS

This chapter discusses in greater detail the competitive patterns
of the markets for Appalachian coal. As discussed in Chapter 2,
Appalachian coal sales can be grouped into several broad categories,
according to purchaser -- electric utility, coke manufacturers, retail
(mainly home heating), process steam and other industrial, and exports.
This chapter focuses primarily on the electric utility market, because
this sector is the largest, fastest growing, and most competitive of
the various markets.

As discussed in Chapter 5, the demand for coal for coking purposes
is quite insensitive to the relative prices of other fuels, depending
mainly on the technology of coke consumption and on steel production.
Relative prices have virtually no perceptible impact on retail coal
sales, coal having been largely displaced from this market by cleaner
and more convenient fuels such as natural gas and distillate fuel oil.
In other words, because such small amounts of Appalachian coal are
currently consumed in this use, differences in decline rate can have
only a negligible effect on total coal consumption.

Consumption by industries in the "other manufacturing and mining" category, largely for the manufacture of process steam, has been, on the whole, stable or declining over the postwar period. Most firms burning coal for this purpose would find it too costly to maintain capacity for burning more than one fuel. This category includes a heterogeneous assortment of industries, so that it is difficult to determine the factors influencing the level of coal consumption. Fluctuations in this category appear to result mainly from changes in the level of general industrial activity. In the short run, there is little apparent effect from changes in relative fuel prices, but, over the period of time needed to replace existing fuel burning equipment, coal use in this category would be more responsive to relative prices.

Exports of U.S. coal depend mainly on the import policies of EEC countries and on the growth in steel production there and in Japan. The relative prices of alternative fuels in the United States influence coal exports only indirectly (through their effect on coal prices).

Fuel consumption in the electric utility market is, on the other hand, highly competitive. Fuel costs are a large proportion of the operating costs of conventional steam-electric plants, and many plants are able to burn more than one kind of fuel, even without additional investment. As discussed in Chapter 2, the costs of converting a coal-burning furnace to an oil-burning one are relatively small (although conversion in the opposite direction can be expensive), and such a change is almost always profitable if the cost per BTU of oil is below that of coal. The choice of burner equipment for a *new* plant is, of course, very sensitive to expected fuel costs. The capital costs of an oil-burning plant are lower than of a coal-burning plant, so that the expected cost per BTU of coal must be *below* that of oil for a coal-burning plant to be chosen.

Several factors influence the cost per BTU to an electric power plant of the various fossil fuels. Prices at the production point, transportation costs, and the federal regulations that govern natural gas are important determinants of the relative fuel costs faced by a utility at a particular location. In the coming decade, the costs of burning different fuels are also likely to be strongly influenced by pollution regulations, especially those pertaining to particulate and sulfur dioxide emissions. Although the growth of nuclear power after 1980 will depend on its competitiveness with fossil fuel alternatives, its largest possible growth by 1980 has already been determined by plans already announced (in view of the seven- to eight-year lead time required). All of these questions are examined in this chapter, and the analysis serves as a base for the demand function for coal developed in Chapter 4.

Prices at Minehead, Wellhead, and Port of Entry

The prices of the fuels at minehead, wellhead, or port of entry account for an important share of their delivered prices. The factors determining these prices are considered separately for each of these three fuels.

35

Coal

Coal prices at the minehead (as distinct from average value per ton) are not readily available.[1] Because of ease of entry into the coal mining industry, in the long run these prices are determined by production costs (including a return on invested capital). As the focus of this study is not on coal production, the primary determinants of mining costs -- mining method, seam characteristics, mine depth and size and, in the short run, the rate of output -- are not discussed in detail here.[2] In general, it appears that current long-run marginal costs lead to a minehead price on the order of $13.00 to $14.00 per ton for high sulfur coal.[3]

Gas

The Federal Power Commission regulates the field price of natural gas sold in interstate commerce. The area pricing system, which places

---

[1]Charles River Associates, *op. cit.*, Volume I, "Profile of the Appalachian Coal Industry and its Competitive Fuels", pp. 247-249, discusses in detail the available data on coal prices.

[2]Other factors also influence the costs at individual mines, but these appear to be the primary ones. *Ibid.*, pp. 319-326.

[3]For example, the FEA estimates that by 1977 coal must sell at $13.50 to $14.00 per ton for new underground mines in Appalachia to open. (Federal Energy Administration, *Project Independence Blueprint: Final Task Force Report on Coal* (Washington 1974), p. 31.) These costs imply a minehead price of about $0.55 per million BTU (*ibid.*, p. 33), or quite close to the figure of $0.40 to $0.50 per million BTU estimated by Richard L. Gordon (*U.S. Coal and the Electric Power Industry, Resources for the Future*, 1975, as cited in Richard L. Gordon, *Economic Analysis of Coal Supply: An Assessment of Existing Studies*, Electric Power Research Institute, May 1975, p. 117). Both sets of figures are in 1973 dollars.

a ceiling on the price received for gas discovered as of a given date in a particular area, has maintained the field price below its market clearing level.[1] This policy has had several effects. First, the low relative market price of natural gas has led to rapid expansion in its demand. Second, because of the low gas price, producers have had less incentive to explore and develop new reserves. Third, because *intra*state gas prices are not regulated by the FPC, this market has grown considerably at the expense of interstate sales. The resulting excess demand for interstate gas has led to rationing, in that many gas distributors are refusing to accept orders for new business.

Partly because of FPC regulation, there is a complex structure of natural gas prices, varying not only by location but by contract date and by type of contract. Table 3-1 shows weighted average rates for new long-term (usually 15 to 20 years) contract sales and for 60-180 day emergency sales. This latter category includes both interstate and intrastate sales, so that it is hard to interpret. It probably reflects market-clearing prices more closely than the area rates do.

---

[1]See, for example, Edward W. Erickson and Robert M. Spann, "Supply Response in a Regulated Industry: The Case of Natural Gas", *Bell Journal of Economics and Management Science* Vol. 2, No. 1 (Spring 1971), pp. 94-121, or Paul W. MacAvoy, "The Regulation Induced Shortage of Natural Gas", *Journal of Law and Economics* (April 1971).

Table 3-1

WEIGHTED AVERAGE NATURAL GAS PRICES,
1971-1975
(Cents Per Thousand Cubic Feet)

Year	New Long-Term Contracts, Area Rates	60-180 Day Emergency Sales
1971	23.81	32.94
1972	24.54	34.43
1973	24.63	49.50
1974	42.85	66.78
1975	52.56	99.73

Source: *FPC News*, Vol. 9, No. 25 (June 18, 1976), pp. 30-31.

It is by no means simple to estimate the market-clearing price of natural gas. The high prices paid for liquefied natural gas and for synthetic gas suggest that the equilibrium price is well above the area rate level shown in Table 3-1.[1] MacAvoy and Pindyck's simulations show that, with "phased deregulation" over the 1972-1980 period, a new contract field price of about $0.90 per Mcf nearly eliminates excess demand by 1980.[2] In any event, natural gas is unlikely to make inroads into coal's share of the electric utility market. As can be seen in Table 2-4, gas's share has declined steadily since 1970.

It seems likely to continue to decline. If its price is kept low by the FPC, then the supply available to electric utilities (which, in areas competitive with coal, often receive gas on an interruptible basis) is likely to diminish rather than grow. Moreover, the FPC has

---

[1]The prices paid for these fuels, used for peaking rather than baseload purposes, are not necessarily the market-clearing price for natural gas. Demand for peak gas may be more inelastic than the demand for baseload gas; it is hard to sort out the role of utility rate regulation in determining the amounts utilities would be willing to pay for natural gas as opposed to capital-intensive coal gasification plants; and the supply elasticity of natural gas, discussed below, also determines the equilibrium price. In spite of these complications, it seems quite certain that the equilibrium price is far above its currently controlled level, especially given the prices of alternative fuels in 1975.

[2]The MacAvoy-Pindyck model is a complex multi-equation model, in which additions to natural gas reserves depend on past prices as well as current ones. Hence, the 1980 price of $0.90 per Mcf depends on a pattern of steadily rising new field prices over the 1972 to 1980 period. Paul W. MacAvoy and Robert S. Pindyck, *The Economics of the Natural Gas Shortage, (1960-1980),* (Amsterdam: North-Holland Publishing Company, 1975).

39

announced its intention to divert the available gas from electric utilities to other users. The courts have thus far ruled that the FPC has the power to do so. The FPC's position, that electric utility consumption is a "low value" use for natural gas, might, alternatively, be interpreted to mean that, if gas was priced sufficiently high to clear the market, electric utilities would no longer buy any.

<u>Oil</u>

Residual fuel oil has been displacing coal from the utility market on the northeastern seaboard only since 1966, when imports of fuel oil were, essentially, granted an exemption from the oil import quota.[1] The price of residual oil in these markets, is, therefore, determined by the world price plus ocean shipping charges. The list prices of residual oil at various East Coast ports, 1969-1975, are shown in Table 3-2. As can be seen in this table, the rise in the price of crude oil since October 1973 is reflected in the 1974 prices for residual fuel oil.

The sharp increase from 1973 to 1974 is discussed in Chapter 7. In addition to the possibility of oil price increases resulting from concerted action by members of the Organization of Petroleum Exporting Countries (OPEC), Venezuela, because of its location, can, within limits, raise its taxes unilaterally.

---

[1]The electric utilities in these states have maintained oil-burning capability throughout the postwar period, but, except for Maine and Florida, they all burned substantial quantities of coal until 1966.

Table 3-2

PRICES OF #6 FUEL OIL (RESIDUAL) EAST COAST PORTS, 1969 TO 1975
(Dollars per Barrel)

Port	Sulfur Content[1]	1969	1970	1971	1972	1973	1974	1975
Baltimore, MD	1%	2.21	2.85	4.12	4.08	4.98	12.07	12.74
Boston, MA	1%	2.21	2.82	4.36	4.20	5.21	12.95	12.80
Charleston, SC	2.1%	2.37	2.85	3.65	3.28	3.92	10.50	11.23
Jacksonville, FL	1.7%	2.35	2.89	3.73	3.23	4.07	10.90	11.80
Miami, FL	NSG	2.33	2.86	3.92	3.76	4.28	10.52	11.21
New Haven, CT	0.5%	2.39	2.88	4.07	4.42	5.33	10.38	11.50
New York, NY	1%	2.33	3.29	4.23	4.08	5.18	12.64	NA
Norfolk, VA	2.1%	2.21	2.75	3.65	3.27	3.92	10.47	11.37
Philadelphia, PA	NSG	2.32	3.16	4.21	4.08	3.89	9.32	NA
Portland, ME	NSG	2.44	3.16	3.86	3.39	4.23	10.93	11.49
Providence, RI	1%	2.43	2.91	4.07	4.20	5.16	13.01	12.80

NOTES: [1]All entries shown are for no sulfur guarantee (NSG) residual fuel oil in 1969; for 1970 to 1974, maximum sulfur content is as indicated in this column.

SOURCE: *Platt's Oil Price Handbook and Oilmanac* (51st edition) (New York: McGraw-Hill, 1975), pp. 51-78. Prices computed from the yearly low and high averages, not from daily averages, all ex-tax. 1975 prices from *ibid.*, (52 edition) (New York: McGraw-Hill, 1976), pp. 44-68.

NA: Price for this sulfur content not available.

There are two main constraints on Venezuela's ability to raise taxes without decreasing total revenue. First, residual fuel oil can be imported from European refineries or more residual fuel oil can be produced at Caribbean refineries. The amount by which the tax could be increased must be less than the sum of the transportation differential and the difference in residual prices at Venezuelan versus Caribbean and European refineries. Because transportation costs are mainly ocean tanker costs, they fluctuate considerably. A second, more general constraint, is the ability of utilities to shift back to coal, although the costs of meeting sulfur emissions standards are so high that other residual fuel oil would be substituted instead.

Because of uncertainty about the future price of residual, I make two alternative assumptions. The first is that the real price of residual fuel oil (no sulfur guarantee) will return to its 1969 level. The second, discussed in greater detail in Chapter 4, is that the price of Venezuelan oil rises close to the lowest of the ceilings discussed above.[1]

Transportation Costs

Coal

Because coal often travels long distances from the minehead to the areas where it is consumed, transportation costs account for a substantial share of the delivered price. Coal moves mainly by water and by

---

[1]The assumptions are based on the case where world crude prices return to their 1969-1970 levels. In the event they remain at their 1975 levels (or higher), the monopoly power of Venezuela is negligible compared to the power of OPEC as a whole, so that the analysis in Chapter 7 of current relative oil is not affected by Venezuela's tax decisions.

rail. If shipments to U.S. ports for overseas exports are excluded, over 50 percent of coal shipments are by rail alone, and another 25 percent move at least part of the way by water (river, tidewater, or Great Lakes). The remainder moves by truck (12 percent) or by tramway, conveyor or private railroad (9 percent).[1]

Rail

The structure of rail rates for coal shipments is very complex, depending on location, quantity of coal, and the distance. From a sample of over 300 rates, most of them for shipments to electric utilities, the cost in cents per ton mile varied between 4.455 and 0.119.3.[2] Although the cost per ton mile tends to be less for greater distances, distance is only one of several important factors.

The rates charged for unit train shipments are, for the purposes of this study, the relevant rates. Introduced in the early 1960's by the railroads to protect coal's share of the electric utility market from nuclear power and oil, and to compete with slurry pipelines, unit trains already accounted in 1974 for about 52 percent of all-rail coal shipments.[3]

---

[1]These estimates, for 1975, were computed from U.S. Bureau of Mines, *Mineral Industry Surveys*, "Bituminous Coal and Lignite Distribution, Quarterly" (Washington: April 1976), pp. 4, 6.

[2]The sample in question was compiled by the U.S. Bureau of Mines, *Transportation Costs of Fossil Fuels*, Interim Report June 1971, Table A-2, pp. 56-59. The rates include the increases under Ex Parte 267A, effective November 1970.

[3]For an excellent analysis of the decision to introduce unit trains, see Paul MacAvoy and James Sloss, *Regulation of Transport Innovation: The ICC and Unit Coal Trains to the East Coast* (New York: 1967); the estimate of unit train shipments was computed from U.S. Bureau of Mines, *Mineral Industry Surveys*, "Coal--Bituminous and Lignite in 1974" (January 27, 1976), pp. 53, 56.

Although complete data are not available, it is quite likely that electric utilities receive a much larger portion of their all-rail coal shipments this way. New, large power plants which receive coal by rail generally have unit train service, and unit trains are often introduced when coal's market share is threatened. In this sense, then, unit train rates represent the marginal cost of transportation for rail shipments to electric utilities.

Unit trains are made up, as a rule, of a set of conventional equipment, used exclusively and virtually continuously to shuttle coal from the mine to the consumption point. Loading and unloading times are minimized, and the equipment is used much more intensively than in ordinary haulage, in which cars spend a large share of their time idle or empty.

To warrant the continuous operation of a 7,000-ton train, of course, both the mine and the consumer must deal in large volumes. For this reason, electric utilities account for most unit train tonnage, although a number of coke and cement plants also receive their coal this way. Most unit train rates specify a minimum annual tonnage, and almost all require a minimum trainload tonnage.

To analyze unit train rates, I fitted a multiple regression by ordinary least squares. The sample consisted of rates from coal mines to electric utilities, with cars owned by the railroad. The data were taken from a survey by the Traffic Department of Peabody Coal Company, prepared in August 1970. The dependent variable was *RATE*, the rate per ton in dollars. The independent variables were *MILES*, the distance in miles; *LOAD*, the average of time allowed for loading and unloading, in hours; and *COAL*, coal's share (in percentage points) of the total

44

fuel burned in 1969 by the generating plant receiving the coal. This
last variable, whose source was *Steam-Electric Plant Factors*, was included
to measure the effect of the availability of alternative fuel supplies
on the railroad rate. There is reason to believe that, the greater
the availability of alternative fuels, the lower the unit train rate,
since a utility's bargaining strength depends on the credibility of its
threats to burn another fuel instead of coal. Thus, this variable is
a crude measure of the utility's ability to burn other fuels -- the
lower the coal's share, the greater the utility's bargaining strength
is assumed to be.

Because time required for loading and unloading ties up equipment
unproductively, it was expected that longer loading times (larger values
of *LOAD)* would be associated with higher rates, when distance and
competitiveness were held constant.

The estimated equation was:

$$\log RATE = \begin{matrix} -3.134 \\ (7.30) \end{matrix} + \begin{matrix} .538 \\ (8.97) \end{matrix} \log MILES + \begin{matrix} .022 \\ (3.95) \end{matrix} LOAD \qquad (1)$$

$$+ \begin{matrix} .150 \\ (1.74) \end{matrix} \log COAL$$

31 observations; $R^2 = 0.836$

Standard error of regression = 0.23

The *t-statistics*, shown in parentheses below the coefficients, are
(except for that of *COAL)* highly significant by conventional statistical
tests. The coefficient of determination indicates that over 80 percent
of the variation in unit train rates is "explained" by the equation.

A number of other variables were tried, including minimum annual tonnage and minimum trainload tonnage. The estimated coefficients of these variables were always either statistically insignificant or qualitatively implausible. Although this negative evidence is open to several interpretations, it seems likely that, once unit train service is established, the cost per ton (and hence the rate) is not influenced by the total tonnage moved.[1]

The sample data were compiled before the rate increase of November 1970; in almost every case the new rate was 14 percent higher than the old one. Thus, the estimated equation, after the rate increase, would be exactly the same as before, except that the constant term should be -3.003 (-3.134 + ln 1.14). For this sample, the average rate per ton (after November 1970) was \$2.91, or about 8 mills per ton mile.

According to the estimated equation, cost increases much less than proportionately to distance, a doubling of mileage leading to only about a 50 percent increase in the rate. On average, a 10 percent increase in the time allowed for loading and unloading means about a 2 percent increase in the rate. As expected, the less competition from other fuels, the higher the unit train rate.

Water

Where the mines and the electric utilities are located close to navigable rivers, barge transportation offers the lowest costs per ton-mile for coal. These costs vary considerably, however, depending

---

[1]This relationship was also found by Martin Zimmerman, *Long Run Mineral Supply: The Case of Coal in the United States*, unpublished Ph.D. dissertation, M.I.T. (September 1975).

on navigational conditions, the number and type of locks, river traffic, and so forth. In particular, if the barges must pass through locks, the locks limit the number of barges in a single tow, and, as the capacity of most locks is far less than the capacity of the rest of the river or canal, the resulting congestion increases the travel time sharply.[1] In 1974, about 66 percent of the tonnage shipped by river traveled on the Monongahela and Ohio rivers.[2]

The costs of moving coal by barge have been estimated to range between 1.7 and 7.2 mills per ton mile, with "typical" costs closer to the lower figure.[3] These costs are considerably below the costs for most unit trains. Indeed, barge costs are so low that in many cases the least cost method of shipment involves a combination of water and rail transportation, even though the costs of transferring the coal from the railroad to the barge or vice versa are on the order of $0.20 to $0.25 per ton.[4] These costs have risen somewhat since the time of the Bureau of Mines survey.

---

[1]A full general discussion can be found in Charles River Associates, Incorporated, *A Study of the Inland Waterway User Charge Program* (Cambridge: December 1970).

[2]U.S. Bureau of Mines, *Mineral Industry Surveys*, "Coal-- Bituminous and Lignite in 1974" (January 27, 1976), p. 52.

[3]These estimates are from Charles J. Johnson, *Coal Demand in the Electric Utility Industry, 1946-1990* (Ph.D. Thesis, The Pennsylvania State University, 1972), Chapter 4, Table I, p. 120.

[4]U.S. Bureau of Mines, *Transportation Costs of Fossil Fuels*, Table A-1, p. 55, no date.

Minemouth Power Generation

An alternative to transporting coal is to locate the generating plant at or near the coal mine and transmit electricity to the consuming regions. This method, called minemouth power generation, substitutes transmission costs for freight costs. On the basis of these costs, minemouth power generation is not very attractive, because it is much more expensive to transmit power than to transport coal -- transmission costs over high voltage lines range between 4 and 10 cents per million BTU per 100 miles, as opposed to a range for unit trains of 1.4 to 3.6.[1]

A number of minemouth plants has been built, however, so that there are other factors that, in some cases, outweigh the transportation cost disadvantages. Air pollution standards may, for example, be much stricter in metropolitan areas than in coal mining areas. In addition, if high voltage interconnecting lines have been constructed to bolster system reliability, and if the mine is located near these lines, then the transmission costs from a minemouth plant are much lower than if the lines have to be put up specifically for that plant.

Other

Costs of moving coal by truck depend considerably on the size of the truck. One set of estimates, constructed for Eastern Kentucky, suggests a range of about $0.15 to $0.10 per ton mile if union labor is employed

---

[1]Johnson, *op. cit.*, Chapter 2, Table 1. The discussion of minemouth power generation follows Johnson's closely.

or about $0.11 to $0.08 with non-union labor.[1] These ranges are for gross weight trucks from 22 tons to 60 tons. These costs, even at their low end, are roughly 10 times the cost of moving coal by unit train. Truck transport is used chiefly to move coal short distances from the minemouth to a rail tipple or to a preparation plant. No information is available on the costs of transporting coal by tramway, conveyor, or private railroad, but these methods are used only for short distances and for a small portion of total coal shipments.

Finally, the slurry pipeline, although of negligible importance at present, should be mentioned. The only one currently in operation carries coal in a slurry of half coal, half water from Black Mesa, Arizona to the Mohave power plant in Southern Nevada.[2] The inhospitable terrain and the lack of a direct rail line appear to have been important factors in the decision to build a pipeline. Unit train service, where the track has already been laid, appears to be less expensive than slurry pipeline. The evidence for this is that the introduction of unit trains in 1963 led to the closing of a pipeline already in operation.[3] Even if train rates rise substantially,

---

[1]Curtis E. Harvey, *Enforcing Weight Restrictions on Eastern Kentucky Roads,* Office of Development Services and Business Research, University of Kentucky, Lexington, Kentucky, p. 86, n.d.

[2]Johnson, *op. cit.,* Chapter 4 ("Transmission"), pp. 128, 131.

[3]*Chemical Week,* March 16, 1968, p. 78.

difficulties in obtaining rights of way are liable to make pipelines generally unattractive, except perhaps in the West.[1]

Oil

Residual fuel has, historically, been transported almost exclusively by water, either in tankers or barges. It can be transported in railroad tank cars and trucks, but it is expensive. In addition to the direct transport costs, the cost and time involved in removing the very viscous oil from the tankers make this method costly. Its viscosity also prevents it from being shipped by conventional pipeline. Heated pipelines have been used in Europe, however, and an insulated pipeline is being planned in Pennsylvania.

Water

The costs of shipping residual fuel oil by tanker vary with distance, size of tanker, and economic conditions in the world tanker market. For shipments along the U.S. coast, the Jones Act raises the costs above what they would be for a similar shipment in the world market. In general, the longer the voyage, the lower the cost per ton mile. Similarly, the larger the capacity of the tanker, the lower the cost per ton mile. During 1975, the average monthly rate for tankers between 80,000 and 159,999 deadweight tons, Ras Tanura to New York, ranged between $1.20

---

[1]See, for example, the account of the inability to gain rights of way from northern West Virginia to four eastern seaboard utilities, given in MacAvoy and Sloss, *op. cit.*, p. 28n.

50

and $1.26 per barrel. The average 1975 rate for large tankers between
Caribbean ports and New York was $0.26 per barrel.[1] Because of the
substantial economies of scale in tanker haulage, costs are higher
for tankers in the 16,000- to 25,000-ton class.

Little, if any, residual oil has been shipped by tanker down the
St. Lawrence Seaway to Great Lakes ports. The trip is somewhat in excess
of 1,000 miles, and, since only small tankers can pass through the locks,
large diseconomies of scale would be incurred. As a rough approximation,
then, costs might be on the order of 5 to 6 mills per ton mile for the
trip itself. Since the Seaway is closed during three winter months,
sufficient inventory would have to be on hand in early December to last
until April, but these costs, under reasonable assumptions, would be small.[2]

At present, domestic shipments of residual fuel oil by tanker are
insignificant. In 1974, for example, shipments from the Gulf Coast
to the East Coast were less than 10 percent of domestic production.[3]
Long distance barge movements were even less significant, with only about
3 percent of domestic production traveling from PAD District III up the
Mississippi River to PAD District II (although substantially more moves
short distances, along the Gulf Intercoastal Waterway, for example).[4]

---

[1] *Platt's Oilgram Price Service and Oilmanac*, 52nd edition (New York:
McGraw-Hill, 1976), pp. 169, 173-174.

[2] For example, at an interest rate of 10 percent per year, the
interest charges on three months' consumption would add less than 1 percent to a year's fuel bill. Costs of storage would be additional.

[3] U.S. Bureau of Mines, *Mineral Industry Surveys*, "Petroleum Statement,
Annual" (April 1, 1976), pp. 18, 29.

[4] *Ibid.*, p. 29. The United States is divided into five Petroleum
Administration for Defense (PAD) Districts. District III includes Texas,
Louisiana, Mississippi, Arkansas, Alabama and New Mexico. District II
includes Ohio, Indiana, Kentucky, Michigan, Tennessee, Illinois, Wisconsin,
Minnesota, Iowa, Missouri, North and South Dakota, Nebraska, Kansas and
Oklahoma.

The lower Mississippi is an uncongested waterway, and barge rates there are quite low, on the order of 1 to 2 mills per ton mile.[1] In the San Francisco Bay Area, on the other hand, the costs are more representative of those prevailing in congested areas, on the order of 12.5 mills per ton mile.[2] Under current conditions, there is no need to move much domestic residual oil any distance, since most of it is produced at refineries close to consuming centers. The discussion here aims not so much at explaining actual movements (which are quite small) as at providing data for the analysis of potential shipments of imported residual fuel oil.

Pipelines

Had OPEC not successfully raised prices in 1973-1974, the costs of sulfur emission control might have induced some utilities away from the coast to substitute residual fuel oil for coal. This possibility is examined in detail in Chapter 6. For such substitution to have occurred, however, in many cases the oil would have had to be shipped by pipeline.

Pipelines in many cases are the least expensive method of shipping petroleum products. Probably about 70 to 80 percent of the total cost of shipping oil by pipeline is the fixed cost of building the pipeline (right of way costs, pipe costs, labor costs, and others) and storage facilities. Operating costs, mainly power costs for pumping stations and maintenance costs, are thus small in relation to the total. Inasmuch

---

[1]Charles River Associates, Incorporated, *A Study of the Inland Waterway User Charge Program*, Chapter 2.

[2]This figure represents the rate from San Francisco to Sacramento (92 miles), as published in San Francisco Barge Tariff Bureau, Local Freight Tariff No. 1, cited in U.S. Bureau of Mines, *Transportation Costs of Fossil Fuels*, p. 28.

as a pipeline built to carry residual oil from a port to an electric utility, say, has virtually no alternative use, the high proportion of fixed costs implies that the utility must expect to burn residual oil over a sufficiently long period to justify the initial expense.

Further, costs per unit decline sharply as pipeline diameter increases, both because capacity increases as the square of the radius and because the internal friction of the oil is less than the friction between the oil and the pipe. Consequently, the utility must be a large volume user if it is to capture the potential low costs of pipelines. For these reasons a utility would want to be sure that residual fuel oil would continue to be the least expensive fuel (in compliance with sulfur emission restrictions) before undertaking the heavy capital costs of pipeline construction.

Although there has been no experience to date in the United States with the pipelining of residual, a number of insulated pipelines have been built in Europe and Australia. Between 1963 and 1968, 11 such pipelines were reported in the literature.[1] They ranged in length from about 4 to 63 miles, and in diameter from 6 to 12 inches.[2] No cost estimates were given, but these pipelines show that insulated pipelines are technically successful, at least for short distances.

---

[1]Jose Gerson Bloch, *Feasibility Study of Offshore Facilities in Massachusetts Bay for Unloading Residual Fuel Oil* (M.S. Thesis, Department of Chemical Engineering, Massachusetts Institute of Technology), March 1971, p. 63.

[2]"Insulated Buried Hot Fuel Oil Pipelines Prove Successful", *Europe and Oil*, Vol. 6, p. 9 (1967), cited in Bloch, *loc. cit.*

In 1969, Pennsylvania Power and Light announced plans to construct

an 80-mile insulated pipeline to carry a mixture of crude oil and

residual oil from Marcus Hook to Martins Creek, Pennsylvania.[1] After

considerable controversy, lawsuits, and delay, the 18-inch pipeline

began operation in August, 1976.[2]

The costs of pipelining crude have been estimated to range

between .23 and .93 mills per barrel mile, with the usual cost about

.35 mills per barrel mile.[3] To take into account costs of heating the

pipeline, higher pumping costs, and perhaps a smaller diameter pipeline

than is usual for crude oil, I assume that the costs of pipelining

---

[1]"Plant Expansion Keyed to Oil Pipeline", *Electrical World*,
October 1, 1971, p. 64.

[2]Press Release from Pennsylvania Power & Light, August 10, 1977.

[3]The range is from Federal Power Commission, *National Power Survey*,
1970, Volume III, p. III-3-118; the "usual" is from *Oil and Gas Journal*,
"Pipeline Installation and Equipment Costs: Oil and Gas Provide Lowest
Energy Transportation Cost", July 29, 1968, p. 116. Costs per barrel-
mile vary considerably depending on pipeline diameter. In 1969, for
example, estimated costs per barrel-mile of pipelines 16 to 24 inches
in diameter were roughly double those of pipelines 30 to 36 inches in
diameter (calculated on the assumption that throughput is roughly
proportional to the square of the pipeline radius, from data on costs
per mile reported to the Federal Power Commission and reproduced in
the *Oil and Gas Journal*, August 4, 1969, pp. 153-154). The range
of costs shown here is broad enough to encompass all of the different
sizes currently being constructed.

residual range between 1 and 2 mills per barrel mile.[1] This range

is largely guesswork, and if more reliable estimates are known, then

they should be substituted.

Gas

Natural gas is also transmitted by pipeline. The costs per mile vary

loosely with distance, ranging from 1 to 2 cents per thousand cubic

feet (Mcf) per hundred miles.[2] Although gas transmission companies cite

the lower figure as a rule-of-thumb estimate, other sources indicate

that 1.5 cents per Mcf per hundred miles is a typical cost.[3] Transmissions

---

[1]It is beyond the scope of this study to estimate an optimal pipeline
network and its characteristics for residual fuel oil. It seems probable,
however, that the oil consumption of even the largest electric utility
plant would be far less than the amount typically carried by a large crude
pipeline operating at economical rates. For residual oil, moreover, the
problems of heat loss and start-up costs suggest that the oil should be
flowing steadily for minimum costs. The case of Pennsylvania Power and
Light, cited above, provides some corroboration for this view, inasmuch
as the planned diameter -- 14 to 16 inches -- is much smaller than the
typical crude oil pipeline diameter.
  The costs assumed here are appropriate for the analysis, as they
reflect the estimate of pipeline costs of utilities in 1970 contemplating
switching from coal to residual fuel oil. Between 1970 and 1975, the
Oil and Gas Journal's pipeline construction cost index rose about 73
percent, although average reported construction costs per mile appear to
have been stable over this period. (*Oil and Gas Journal,* August 23,
1976, pp. 83-84.)

[2]U.S. Bureau of Mines, *Transportation Costs for Fossil Fuels,*
p. 40.

[3]The latter figure is cited in *Oil and Gas Journal,* July 29, 1968,
p. 116, and it is also the average for the pipelining systems examined
in *Transportation Costs for Fossil Fuels.*

costs thus increase the delivered price of natural gas substantially, at least in areas far from the producing fields.

Distribution costs are another substantial addition for most users, but not for electric utilities as a rule. For those electric utilities that also distribute natural gas, burning the gas themselves clearly involves no additional distribution costs. Other electric utilities, especially in the areas far from the producing fields, buy it at off-peak times (such as the summer months). As large-volume users, they pay much the same rate as the gas distributing companies themselves.[1] Transportation costs for natural gas are not critical to this study, however, because gas is not likely to encroach further on the markets for Appalachian coal.

Patterns of Coal Shipments

The prices and costs discussed above determine which fuel will be the cheapest source of BTU's at any given generating plant.[2] In the short run, of course, the costs may differ considerably from those discussed. For example, a sharp increase in coal prices might cause coal to be more expensive than imported residual oil pipelined into central New York State. Unless these relative prices were expected to obtain for the next 10 or 15 years, however, a utility would prefer to burn the (temporarily) high-priced coal rather than undertake the construction of a pipeline. In other words, there may be large adjustment costs

---

[1]U.S. Bureau of Mines, *Transportation Costs for Fossil Fuels*, p. 44.

[2]Although the discussion has been in terms of physical units -- tons, barrels, and cubic feet -- conversion to BTU equivalents is straightforward.

associated with a change in fuels; consequently, not only will the pattern prevailing at a particular time reflect past decisions and market conditions as well as current ones, but current decisions will be based on expected as well as current market conditions.

Data on continental coal shipments are published showing origin by producing district, destination by state, major end use, and mode of transport. These data show that most Appalachian coal sold to utilities moves relatively short distances. None, for example, moves west of the Mississippi: in the West South Central states, utilities burn natural gas almost exclusively; the West North Central states either use gas or Western coal.[1] Very little coal moves into New England, where the distance from the coal fields makes coal expensive and where air quality regulations have required the use of low sulfur fuels. Appalachian coal shipments to New York and Pennsylvania are almost exclusively to

---

[1]Rising gas prices (especially in unregulated intrastate sales), coupled with the unavailability of gas as a boiler fuel, have recently led some utilities in the West South Central states to start burning low sulfur Western coal. Although the amounts are not large, coal is also being planned for a large fraction of future fossil fuel consumption by utilities in these states. For example, in 1969 utilities in Texas burned no coal, while in 1974 they burned about 5 million tons (or about 5 percent of total BTU's consumed). *(Steam-Electric Plant Factors*, 1970 and 1975 editions, Table 2). Coal accounted for 3 percent of total BTU's in 1974 in the West South Central region, but, of the additional fossil capacity scheduled to come on-stream between 1974 and 1979, over 40 percent is expected to be coal-burning. Moreover, if coal's share is disaggregated by year, it is seen to be rising over this period: 1974 and 1975 - 11 percent; 1976 -- 25 percent; 1977 -- 58 percent; 1978 -- 84 percent; 1979 -- 78 percent *(Steam-Electric Plant Factors*, 1974 edition, Table 4).

the interiors of those states, utilities on the coast burning residual fuel because of its lower cost and because of sulfur restrictions. Appalachian coal still has a competitive edge in the Appalachian states themselves and in the adjacent South Atlantic states, where closeness to the coal fields results in a low delivered price.

Delivered Prices

Transportation costs account for a significant share of delivered prices, but this share varies among the fuels and among different consuming areas. It would be tedious and to little purpose to estimate this share for each fuel at every location, but it might be useful to compute some broad ranges.

Coal

It is particularly hard to make such estimates for coal, because there are no reliable price quotations, either at the mine or delivered.[1] As mentioned earlier in this chapter, the current price of Appalachian coal at the minehead is around $13 to $14 per ton. Published unit train rates fall, for the most part, in the range of $1 to $6 per ton, with the average around $3.[2] In Table 3-3 are shown, for illustrative purposes, the share of delivered prices represented by transportation costs for various values in these ranges. The share of transportation

---

[1]The problems with the available price series are discussed in Charles River Associates, *op. cit.*, pp. 247-249.

[2]This average includes the subsequent rate increase in November, 1970. In a few exceptional cases, the rates are higher than $6 per ton, but they do not apply to Appalachian utility coal. U.S. Bureau of Mines, *Transportation Costs of Fossil Fuels*, Table A-2, pp. 56-59.

Table 3-3

SHARE OF TRANSPORT COSTS IN DELIVERED PRICES,
FOR VARIOUS MINEHEAD PRICES AND RAIL RATES

Unit Train Rates, Dollars per Ton	Dollars per Ton, F.O.B. Mines		
	$13	$13.50	$14.00
$ 1	8%	7%	7%
$ 3	23%	22%	21%
$ 6	46%	44%	43%

NOTE: These figures are believed to bracket most utility coal
prices and most unit train rates to utilities. A
fuller discussion is contained in the text.

59

costs in the delivered price can range from 7 to 46 percent, although the typical share is probably about 22 percent.[1]

Oil

For utilities at port cities, ocean shipping charges represent the main transport costs of residual fuel oil. Ocean shipping rates fluctuate substantially, but transport costs per barrel of imported oil in 1975 ranged, say, between $0.26 and $1.26 per barrel. At delivered prices of $12 to $13 per barrel, these charges account for betweeen 2 and 10 percent of total costs.

For those utilities inland, reached by truck or by barge, there are additional transportation costs. I am unaware of direct estimates of these costs, but a comparison of delivered prices at inland utilities with the costs at port cities suggests that they are not substantial. For example, the average oil cost at Middletown, Connecticut in 1974 was only $0.64 higher than that at Bridgeport, while the cost at Hartford, Connecticut was $1.51 above that at Bridgeport.[2] (It is difficult to compare costs at Albany or other upstate New York sites with port costs because of differences in sulfur content of the oil.) Inclusive of ocean shipping charges, then, transportation costs to nearby inland utilities account for roughly 7 to 21 percent of delivered costs.

---

[1]The "typical" estimate assumes transport charges of $3 per ton and coal costs of $13.50 per ton at the mine.

[2]*Steam-Electric Plant Factors*, 1975 edition, Table 1.

Gas

Because of the lack of origin and destination data, it is difficult to determine the longest distance that natural gas is currently transported. For gas consumed close to the producing region, transportation costs are a small share of the delivered price, probably on the order of 10 to 15 percent.[1] For shipments of about 2,000 miles on the other hand, such as contracts for the delivery of gas from Texas and Louisiana to the New York area, the share probably ranges from 50 to 75 percent, or about $0.20 to $0.30 out of a delivered cost of $0.40 to $0.44.[2]

Because of these differences in the delivered prices of fuels among various regions, choice of fuel by electric utilities in different sections of the country varies systematically. Table 3-4, which shows the percentage of BTU's supplied by each of the three fossil fuels, by state, summarizes these differences. For purposes of comparison, Table 3-5 shows the same information for 1973. It shows much the same pattern as Table 3-4, with a few exceptions (coal being phased out of New England and some South Atlantic states, some oil being burned in Michigan, some growth in coal use in the West South Central and Mountain states).

---

[1]Based on a wellhead price of $0.14 to $0.20 per Mcf, and transportation costs of $0.01 to $0.015 per Mcf per 100 miles. Because wellhead prices have risen sharply since 1970, and pipeline charges have not increased at the same rate, the share of transportation in the delivered price is considerably lower than stated in the text. For destinations close to the producing regions, transportation costs are now probably less than 5 percent of the delivered price; for shipments of about 2,000 miles, transportation charges probably account for between 25 and 30 percent.

[2]These contracts are discussed in *Transportation Costs of Fossil Fuels*, pp. 39-43, which is the source for the pipeline charges, while average wellhead value in the producing areas in 1970 was taken as the wellhead price.

Table 3-4

FUEL CONSUMPTION PATTERNS OF ELECTRIC UTILITIES BY STATE, 1969

State, by Region	Coal (Millions of tons)	Oil (Millions of Barrels)	Gas (Billions of Cubic Feet)	Percent Consumption in BTU's		
				Coal	Oil	Gas
New England						
Connecticut	2.1	17.8	0.1	32	68	-
Maine	-	4.8	-	-	100	-
Massachusetts	1.9	31.5	5.1	19	79	2
New Hampshire	0.9	2.0	-	67	33	-
Rhode Island	-	3.0	1.1	-	94	6
Vermont	(*)	(*)	-	93	7	-
Middle Atlantic						
New Jersey	4.3	31.9	35.9	32	57	11
New York State	12.9	42.6	106.3	47	37	16
New York City	3.9	42.1	90.0	23	57	20
New York (excl. N.Y.C.)	9.0	0.5	15.4	92	1	7
Pennsylvania	26.5	19.8	5.2	83	16	1
Philadelphia	3.5	19.5	4.6	41	57	2
PA (excl. Phil.)	23.0	0.3	0.6	100	-	-
E. N. Central						
Illinois	29.7	0.3	72.1	90	-	10
Indiana	22.6	0.1	18.7	96	-	4
Michigan	20.3	1.2	27.5	94	1	5
Ohio	33.9	0.3	11.7	99	-	1
Wisconsin	9.4	0.2	26.2	89	-	11
W. N. Central						
Iowa	3.7	0.1	64.0	55	-	45
Kansas	0.2	0.2	151.4	5	1	94
Minnesota	5.2	0.6	59.5	64	2	34
Missouri	8.9	0.1	57.2	78	-	22
Nebraska	0.9	0.1	41.6	34	1	65
North Dakota	2.9	(*)	(*)	100	-	-
South Dakota	0.3	0.1	2.9	59	4	37
South Atlantic						
Delaware	1.3	0.9	3.0	80	13	7
District of Columbia	0.8	2.0	-	63	37	-
Florida	4.5	35.6	176.2	21	44	35
Georgia	7.5	0.5	34.8	83	1	16

Table 3-4 (Continued)

FUEL CONSUMPTION PATTERNS OF ELECTRIC UTILITIES BY STATE, 1969

State, by Region	Coal (Millions of tons)	Oil (Millions of Barrels)	Gas (Billions of Cubic Feet)	Percent Consumption in BTU's		
				Coal	Oil	Gas
Maryland	6.8	4.0	0.1	88	12	-
North Carolina	16.2	0.1	4.8	99	-	1
South Carolina	3.6	0.9	27.9	73	9	23
Virginia	7.9	8.7	1.8	78	21	1
West Virginia	14.6	(*)	0.8	100	-	-
E. S. Central						
Alabama	15.7	-	15.3	96	-	4
Kentucky	14.5	(*)	6.8	98	-	2
Mississippi	0.5	0.3	83.8	13	2	85
Tennessee	15.1	-	19.3	94	-	6
W. S. Central						
Arkansas	-	0.3	83.7	-	2	98
Louisiana	-	(*)	301.0	-	-	100
Oklahoma	(*)	(*)	188.9	-	-	100
Texas	(*)	(*)	956.0	-	-	100
Mountain						
Arizona	0.4	(*)	56.0	12	1	87
Colorado	2.9	0.2	46.7	59	1	40
Montana	0.6	0.1	1.5	78	6	16
Nevada	0.6	(*)	19.7	43	1	56
New Mexico	2.8	(*)	53.6	47	-	53
Utah	0.4	1.6	3.3	41	46	13
Wyoming	3.1	(*)	2.3	95	-	5
Pacific						
California	-	20.9	584.2	-	17	83
Oregon	-	(*)	0.4	-	20	80
Washington	-	(*)	-	-	100	-
U.S. TOTAL	306.4	232.6	3358.3	59	12	27

(*) = less than 0.05 units.

SOURCE: National Coal Association, *Steam-Electric Plant Factors*, 1970 edition, pp. 51-52.

Table 3-5

FUEL CONSUMPTION PATTERNS OF ELECTRIC UTILITIES BY STATE, 1973

State, by Region	Coal (Millions of Tons)	Oil (Millions of Barrels)	Gas (Billions of Cubic Feet)	Percent Consumption in BTU's		
				Coal	Oil	Gas
New England						
Connecticut	(*)	29.3	-	-	100	-
Maine	-	4.9	-	-	100	-
Massachusetts	(*)	44.8	4.7	-	98	2
New Hampshire	1.0	2.0	-	69	31	-
Rhode Island	-	2.5	(*)	-	100	-
Vermont	(*)	(*)	0.5	53	2	45
Middle Atlantic						
New Jersey	2.4	37.5	17.8	19	75	6
New York State	5.8	82.6	46.9	20	73	7
New York City	-	61.6	36.6	-	91	9
New York (excl. N.Y.C.)	5.8	21.0	10.4	50	46	4
Pennsylvania	38.9	17.5	0.4	90	10	-
Philadelphia	2.3	16.5	-	37	63	-
PA (excl. Phil.)	36.6	1.0	0.4	99	1	-
E. N. Central						
Illinois	32.5	7.2	23.5	91	6	3
Indiana	26.9	0.8	6.0	98	1	1
Michigan	20.5	11.8	28.2	83	12	5
Ohio	43.5	1.8	9.6	98	1	1
Wisconsin	10.1	0.9	18.6	90	2	8
W. N. Central						
Iowa	5.3	0.1	54.6	67	0	33
Kansas	1.0	0.8	160.2	12	3	85
Minnesota	6.9	1.0	49.6	70	3	27
Missouri	15.6	0.4	50.9	86	1	13
Nebraska	1.3	0.1	49.3	38	1	61
North Dakota	4.8	(*)	(*)	100	-	-
South Dakota	0.4	0.2	3.6	56	10	34
South Atlantic						
Delaware	0.9	5.6	2.1	37	59	4
District of Columbia	0.2	5.3	-	16	84	-
Florida	6.6	67.4	147.0	21	58	21
Georgia	10.9	3.9	29.7	82	8	10

Table 3-5 (Continued)
FUEL CONSUMPTION PATTERNS OF ELECTRIC UTILITIES BY STATE, 1973

State, by Region	Coal (Millions of Tons)	Oil (Millions of Barrels)	Gas (Billions of Cubic Feet)	Percent Consumption in BTU's		
				Coal	Oil	Gas
Maryland	3.9	24.2	–	39	61	–
North Carolina	19.9	5.2	1.7	94	6	–
South Carolina	5.5	3.5	18.4	77	12	11
Virginia	4.9	24.5	1.5	44	55	1
West Virginia	22.8	0.5	(*)	99	1	–
E. S. Central						
Alabama	18.2	(*)	2.4	99	–	1
Kentucky	22.4	0.2	7.9	98	–	2
Mississippi	1.2	5.7	53.0	24	30	46
Tennessee	17.4	–	11.0	97	–	3
W. S. Central						
Arkansas	–	7.1	48.6	–	48	52
Louisiana	–	7.6	339.0	–	11	89
Oklahoma	(*)	0.1	263.1	–	–	100
Texas	4.7	6.3	1248.8	5	3	92
Mountain						
Arizona	0.5	6.9	45.2	10	42	48
Colorado	4.6	0.5	56.4	65	2	33
Montana	0.9	0.1	0.9	88	5	7
Nevada	3.9	0.3	38.4	67	1	32
New Mexico	7.5	0.8	63.5	66	2	32
Utah	1.0	0.3	4.0	80	7	13
Wyoming	5.8	0.1	0.4	99	1	–
Pacific						
California	–	73.8	452.4	–	49	51
Oregon	–	–	1.1	–	–	100
Washington	3.7	0.4	–	95	5	–
U.S. TOTAL	384.6	496.6	3361.2	57	20	23

(*) = less than 0.05 units.

SOURCE: National Coal Association, *Steam-Electric Plant Factors*, 1974 edition, pp. 53-54.

Because both oil and gas are occasionally used for peaking purposes, and because of the use of interruptible gas, most coal-burning states also use some quantities of the competitive fuels. The patterns would also stand out more clearly if the seaboard portions of East Coast states were disaggregated from their inland portions.

Chapter 4

ELECTRIC UTILITY DEMAND FOR APPALACHIAN COAL

Introduction

In this chapter and the next, I analyze competitive patterns in the markets for Appalachian coal. Although Appalachian coal is used for a number of other purposes (including coking, steam, and residential heating), electric utility consumption has been at once the largest, fastest growing, and most sensitive to sulfur emission standards and to the price of coal relative to that of residual fuel oil.

This chapter, therefore, focuses on coal used by electric utilities for power generation. The first section presents an extrapolation of coal use in this market, on the assumption that no plants burning coal or planning to burn coal will convert to an alternative fuel. This extrapolation is based on electricity consumption growth rates prior to the sharp oil price increases of 1973-1974. It thus provides an estimate of what utility consumption of Appalachian coal might have been in the absence either of changes in relative fuel prices or of a change in the growth rate of electricity consumption.

The second section modifies this extrapolation by assuming that coal-burning plants convert to fuel oil when it is economical to do so. Two other assumptions of this section are that relative fuel prices are at their 1969 level, and that no sulfur emission standards are in effect. This forecast thus measures utility consumption of Appalachian coal at the old relative prices and under the lack of a clean air policy. In Chapter 6 I analyze the impact of alternative sulfur emission standards, while in Chapter 7 I analyze the impact of the change in relative prices of fuel oil and coal after 1973-1974. Chapter 5 is devoted to the demand for coking coal, coal for exports, coal used by "other mining and manufacturing" industries, and retail consumption, including a forecast of coal consumption in each use.

Estimates are also made of the share of Appalachian coal in each category. The benchmark forecast for total Appalachian coal use is the sum of Appalachian coal use forecasted in individual sectors.

Extrapolation of Electric Utility Demand

In 1975, electric utilities accounted for almost 70 percent of U.S. coal consumption, and almost 80 percent of steam coal consumption. Utilities consumed about 56 percent of Appalachian coal, and about 71 percent of Appalachian steam coal.[1] These percentages have been gradually increasing in recent years, reflecting the growing importance of utility coal consumption.

---

[1] In a strict sense, these percentages are based on shipments received rather than consumption, but they should be quite representative of consumption. Consumption figures for 1975 were not available as of the time of this writing. U.S. Bureau of Mines, *Mineral Industry Surveys*, "Bituminous Coal and Lignite Distribution, Calendar Year 1975", April 12, 1976, pp. 4-5.

Prior to the 1973 embargo, electric power generation grew steadily at a rate of about 7 percent per year. If it had continued to grow at that rate, and if coal retained its market share, coal use in this sector would also grow at that rate. As discussed in Chapter 5, no other domestic consumption sector shows signs of substantial growth. Exports may, perhaps, increase in percentage terms, but the impact on total production will be relatively small. Changes in electric utility consumption, therefore, will largely determine whether or not Appalachian coal output increases during the 1970's and 1980's.

There are two potential obstacles to a steady increase in coal consumption by the electric utilities. First, increases in nuclear generating capacity in coal-burning regions decrease the market for coal. Second, strictly enforced sulfur oxide emission regulations may lead utilities to substitute low sulfur residual fuel oil or low sulfur Western coal for Appalachian coal. These two questions pose quite different problems for the analysis of coal demand during the 1970's.

As discussed in Chapter 2, the average delay between the announcement of a new nuclear plant and the start of its commercial operation is, at present, around nine or ten years. It is safe to assume, consequently, that only plants that have already been announced will be in operation by 1980. In contrast, conventional fossil fuel plants have a lead time of only five or six years, so that it is a reasonable assumption that expansion of conventional capacity will occur as needed during the 1970's. There may be some delays due to siting difficulties, but such delays will mean merely that older, less efficient plants will be used more intensively, with slight increases in fuel consumption above the level implied by the assumption.

On the other hand, the expansion of nuclear capacity may lag behind current forecasts, the result, perhaps, of environmental or safety controversies or of construction difficulties. Although the technology of "conventional" nuclear reactors is well-developed, delays may result from a number of other circumstances. Unexpected delays in the past have frequently thwarted predictions of a rapid expansion in nuclear capacity, and such a possibility should be considered for the future.

This section derives forecasts for coal demanded by electric utilities, given expected installations of additional nuclear capacity. This forecast assumes both that no coal-burning plants switch to residual fuel oil, and that no oil-burning plants switch to coal. This extrapolation assumes, in effect, that the relative shares of coal and oil do not change and that electricity consumption grows at its pre-embargo rate.

In the next section, I estimate how much coal would be burned by utilities if, where economical, they switch to oil from coal (assuming 1969 relative prices). This forecast is based on the assumption that such switches occur only in response to the relative prices of oil and coal (and not, for example, in response to sulfur emission requirements), and that these relative prices are at their pre-embargo level. As stated above, this forecast thus provides a baseline for consumption in the absence of both environmental restrictions and OPEC's oil price increases.

Although natural gas is burned by some utilities, it is largely ignored in this analysis. As discussed in Chapter 3, the regulation-induced shortage of natural gas prevents it from displacing either coal or residual fuel. In some areas, indeed, coal is displacing natural gas, but the coal is low sulfur Western coal. Thus, from the standpoint of Appalachian coal, natural gas can be largely ignored.

70

The method I use to extrapolate utility coal use to 1980 is as follows. First, I estimate 1980 net generation, by state, including both nuclear and fossil plants. Second, I subtract from this total an estimate of nuclear generation. This yields the net growth in fossil generation. Third, I estimate the amount of coal that would be burned in 1980 by existing plants and by additions already scheduled to come on-stream by 1973. Finally, I adjust coal use from the previous step to take account of probable but as yet unscheduled additions to coal-fired capacity between 1973 and 1980, using the fossil-fired generation data from the second step.

Other forecasting methods could be used. This method takes advantage of the long nuclear lead times to derive the growth in fossil fuel consumption. Its purpose is to provide a base forecast of utility coal use by state. Its main weakness is in the extrapolation of net electricity generation, but it was beyond the scope of this study to develop a more sophisticated model of electricity generation.

Net Electricity Generation

Because this study focuses on Appalachian coal, the only states considered were those in which some Appalachian coal was burned by electric utilities in 1970. There were 23 such states, plus the District of Columbia. It is frequently convenient to refer to groups of states by their census region (the states considered here belong to the New England, Middle Atlantic, South Atlantic, East North Central and East South Central regions), although not every state in each region is included in the sample.

71

Rather than estimating a structural equation for electricity generation, I used a growth rate based on a simple time trend. For each of the eight Federal Power Commission regions, Johnson fitted semi-logarithmic trend lines to generation data for 1961 to 1973.[1] This procedure yielded rather good fits by conventional statistical criteria, and so his implied growth *rates* were used to extrapolate net power generation for each state.[2] For example, I assumed that the growth rate for FPC Region I (which includes New England and most of the Middle Atlantic states) would describe the trend in power generation in Massachusetts, Connecticut, New York, Pennsylvania, etc. Although this is only an approximation as far as individual states are concerned, it should be fairly accurate. At the regional level, it would be even more accurate.

Growth in Nuclear Capacity

Nuclear capacity in 1980 was assumed to be the sum of nuclear capacity in 1970 plus some of the capacity planned to be installed between 1971 and 1980. I derived nuclear capacity for each state from data on individual plants.[3]

---

[1]Charles J. Johnson, *Coal Demand in the Electric Utility Industry 1946-1990* (Ph.D. thesis, The Pennsylvania State University, 1972), pp. 202, 208. Actual generation figures were used for 1961 to 1970. The 1971-1973 figures were from FPC forecasts. Inclusion of the FPC forecast years improved the statistical properties of the estimated equations, according to Johnson.

[2]Because the FPC regions do not coincide exactly with state lines, states in more than one region were assigned to that region having the larger geographic portion of the states in question. Most states presented no difficulties.

[3]Southern Nuclear Engineering, Inc., *A Listing of Commercial Nuclear Power Plants*, Edition No. 4, October 1971; also reported in *Electrical World*, October 15, 1971, p. 40.

As mentioned, I had to make an assumption as to how much of the planned capacity would in fact be operating by 1980. Although a variety of assumptions might be tried, I assumed that capacity planned for 1970 to 1972 would be installed as planned, but, in each subsequent two-year period, additions would be delayed by one year. This progressive slippage implies, for example, that 1973 installations would be the same as those planned for 1972, 1975 installations those planned for 1973, and 1980 installations those planned for 1976. This assumption was chosen to represent a situation where delays snowball, becoming longer for plants scheduled in the more distant future.

It would be unreasonable to extrapolate this assumption indefinitely, because electric utility planners would revise their estimates in the light of actual construction experience. For the relatively short period considered here, however, it is a reasonable assumption. The results are little changed, for instance, if it is assumed that no slippage occurs.

I should again point out that I have not tried to model the choice between nuclear and fossil plants. It might be the case, for example, that fossil plants are being built only as an interim measure until nuclear plants can be built. In view of the long plant lives relative to the difference in construction times, the fact that large new fossil plants are being planned makes this possibility unlikely. This study, however, does not directly address the question of utility fuel choice after 1980.

To translate nuclear capacity into nuclear generation, I assumed that the plants operate at a 70 percent load factor. This load factor was the highest observed for nuclear plants operating in 1969 (the average was about 60 percent), but it is a typical load factor for large plants used for base-load generation. I then subtracted nuclear net generation from total net generation to obtain net generation by fossil fuels.

Coal Consumption by Existing Plants and Announced Additions

Three categories of coal-fired generating capacity were included in the calculation of 1980 coal use, and because of slight differences in treatment each is discussed separately. The first category is all of the plants that burned 0.05 million tons of coal or more in 1969. The second category includes new plants that will burn coal by 1973. A third category, actually a subset of the first, includes existing plants that will add coal-fired capacity by 1973. The year 1973 was chosen as the cut-off because, given the three-to-four-year delay between the time a large coal-burning power plant is announced and the time it comes on-stream, it seemed quite unlikely that any as-yet unannounced plants could be added before then.

Existing Power Plants

Coal use in 1980 by power plants existing in 1969 was assumed to be 80 percent of the 1969 level. This downward adjustment was made for two reasons, both of which reflect efforts by electric utilities to generate a given amount of power as cheaply as possible. In general, newer plants tend to use less coal per kilowatt hour (in industry parlance, have lower heat rates) and hence to have lower operating costs than older plants. Consequently, older plants are, as a rule, run at lower load factors than newer plants. I assumed a downward adjustment of 10 percent to reflect this decline in the load factor.

74

In addition, however, this decrease in utilization does not apply uniformly to *all* the 1969 plants. Rather, the oldest plants are used disproportionately less, some perhaps being retired, others being placed on stand-by, and so forth. Because the oldest plants use more coal per kilowatt hour, I subtracted an additional 10 percent from 1969 coal use to reflect the shift in generation toward the newer 1969 plants.

Both of these adjustments are somewhat arbitrary, and they probably overstate somewhat the decrease in coal burned at existing plants. It is, however, difficult to verify the assumption directly, because the consumption data are on a plant basis, but individual units are brought on-stream and retired from plants.

A simple check of 35 plants in New York, Pennsylvania, Michigan and Ohio that burned 100 percent coal in both 1964 and 1974 and whose rated capacity did not increase over that period showed a decrease of only 6 percent of coal consumption (measured in heat units).[1] In view of the substantial projected growth in generating capacity over the forecast period, however, the discrepancy is unlikely to affect the results appreciably.

New Power Plants

The calculation of coal used at new coal-fired plants was somewhat simpler. Since new plants tend to be base-loaded (that is, used steadily except for required maintenance) a load factor of 70 percent was applied to total capacity expected to be in place by 1973. To translate this estimate of net generation into an estimate of coal use, a heat rate of

---

[1] Data were taken from National Coal Association, *Steam-Electric Plant Factors*, 1964 and 1975 editions. The plants ranged in capacity from 60 to 10863 megawatts.

9,000 BTU's per kilowatt hour was assumed. This heat rate, *slightly above the level of the best coal-fired plants reached in 1969,* seems plausible, because the minimum heat rate has been leveling off in recent years and, given current technology, will probably improve only marginally by 1980. Finally, the BTU content of the coal burned by the new plant was assumed to be at the 1969 average value for the plant's home state. This choice implicitly assumes that the coal will come from sources similar to those supplying neighboring electric utilities.

For example, 1980 coal use by a power plant brought into operation in 1976 can be written in symbols as:

$$q(J, 1980) = \frac{9000 \times 6.132 \times X(J,1976)}{B} \tag{2}$$

where:

$X(J,1976)$ is the rated plant capacity in thousands of kilowatts;

6.132 converts thousands of kilowatts to millions of kilowatt hours (at a 70 percent load factor);

9000 is BTU's per kilowatt hour; and

B is millions of BTU's per ton of coal.

Old Plants Adding Capacity

New units are frequently added to existing plants. Where such additions were planned, the plants were treated as a composite of the two previous categories. That is, the 1980 coal burn was broken down into coal burned by the existing 1969 plant and coal burned by the planned units. Coal burned by the 1969 capacity was computed according to the formula for existing power plants, while coal used by the planned new unit or units was computed according to the formula for new power plants.

76

Adjustment for Unannounced New Capacity

The procedure outlined above does not take into account that, between 1973 (the last date at which planned additions were reported, as of 1969) and 1980, other new coal-fired capacity may be added (if it is the least-cost fossil fuel). An adjustment was therefore made on this account. Estimated 1973 coal use, based on existing and planned plants, was taken as the base.

The estimated growth rate of fossil-fueled generation from 1973 to 1980 was multiplied by 1973 coal use at each plant to obtain an estimate of 1980 coal use. This adjustment implies, for the model of utility coal demand discussed later in this chapter, that the geographical distribution of coal demand in 1980, in the absence of further interfuel competition, is proportional to that planned and in operation as of 1973.

In symbols, the adjustment, for all the plants, can be written as:

$$q(J,1980) = q(J,1973) \cdot \frac{G(S_J,1980)}{G(S_J,1973)} \tag{3}$$

where $G(S_J,year)$ is fossil-fuel generation in state S_J (the state in which utility J is located) estimated according to the procedure described above. This adjustment factor is applied both to existing power plants (whose coal use was adjusted downward from the 1969 level, as described above) and to plants put into operation between 1969 and 1973.

The model does not treat minemouth power generation separately. That is, for plants whose sites are known (either because the plant has already been built or because its site has been announced), coal and oil transportation costs have been accounted for. If, however, as-yet-unsited capacity shifts dramatically toward minemouth locations, then the assumption about

77

the location of new capacity will be wrong. The probable effects of such
a shift would be to increase coal use above that estimated.

There are two main reasons for this neglect. First, the task of
estimating new plant locations lies beyond the scope of this study.
Second, even if plant location could be estimated, the resulting gain
in accuracy of analysis would probably be small. Since there have been
no recent trends either toward or away from minemouth location of plants,
it is probable that, on average, additions to generating capacity will
conform fairly closely to past patterns of location. That is, if loads
grow more or less at the same rate within any electric utility's generating region, the same geographical distribution of capacity will also
continue to be chosen, other things equal.

Air pollution standards will not upset this tendency. Although
metropolitan areas generally do have stricter air pollution standards
than mining areas, the sulfur-in-fuel standard of 0.7 percent that is
analyzed in this study will, by 1975, be a minimal requirement throughout
most of the United States. Enforcement to date has been very lax, in
part perhaps due to the high costs of meeting the standards, but it seems
unlikely that utility planners will count on the standards remaining
unenforced indefinitely.

It is expected that most coal-fired plants will have to meet this
standard by 1980.[1] Hence the pollution standard provides little incentive
to locate plants away from the consuming regions. Further, overhead

---

[1]For example, one estimate is that almost 60 percent of all coal-
fired plant capacity will have to meet roughly a 0.7 percent standard
by 1980. (Environmental Protection Agency, *Report to Congress on Control
of Sulfur Oxides* (February 1975), p. 15.

transmission lines have stirred considerable opposition from environ-
mentalists, and it is increasingly difficult for utilities to obtain
rights of way for high voltage overhead lines. Putting such lines
underground is prohibitively expensive, so that, all in all, it seems
quite unlikely that minemouth power generation will affect the results
of the model in any important way.

Results

The resulting forecast of coal use, on the assumption that no markets
are lost to competing fossil fuels, appears in Table 4-1. It shows
estimated utility coal consumption by state and region. Table 4-2
distributes this coal among the Appalachian producing districts, according
to the 1970 pattern of shipments to each state.

As can be seen in Table 4-1, these forecasts imply widely varying
growth rates among states. Growth in coal use, as forecasted here,
depends on several factors -- growth in total electric power generation,
growth in nuclear power generation, and growth in planned coal-fired
capacity. In the New England and Middle Atlantic states, no new coal-
burning plants are planned, and these areas have a relatively high
expected growth in nuclear generation. Table 4-3 summarizes the forecasted
growth rates of total generation. Most of the new coal-fired plants
are planned for the East Central and South Atlantic regions, and total
generation in these areas has grown at faster rates than in the New
England and Middle Atlantic regions. Thus, the biggest expected gain
in coal use is in the East North Central region. As of 1970, utilities
in Tennessee, South Carolina and Virginia all expected to add sizeable
amounts of nuclear generating capacity, leading to a slow growth in

Table 4-1

BASE FORECAST OF 1980 COAL USE BY ELECTRIC UTILITIES,
ASSUMING NO MARKETS LOST TO RESIDUAL FUEL OIL
(Millions of Tons)

State, by Region	1969 Coal Use	1980 Forecast	Implied Growth Rate (Percent)[1]
Connecticut	2.11	2.33	0.90
Massachusetts	1.88	2.49	2.55
New Hampshire	0.93	1.17	2.09
New England Total	4.93	5.99	1.77
New Jersey	4.29	3.80	-1.10
New York	12.85	14.73	1.24
Pennsylvania	26.31	41.72	4.19
Middle Atlantic Total	43.44	60.25	2.97
Illinois	29.53	60.46	6.51
Indiana	22.53	42.12	5.69
Michigan	19.89	35.31	5.22
Ohio	33.53	60.58	5.38
Wisconsin	9.24	14.78	4 27
East North Central Total	114.72	213.26	5.64
Delaware	1.34	2.24	4.67
District of Columbia	0.78	0.48	-4.41
Florida	4.53	8.11	5.29
Georgia	7.49	13.31	5.23
Maryland	6.76	11.87	5.12
North Carolina	16.21	20.24	2.02
South Carolina	3.50	4.13	1.50
Virginia	7.89	8.48	0.66
West Virginia	14.55	41.50	9.53
South Atlantic Total	63.05	110.36	5.09
Alabama	15.60	22.58	3.36
Kentucky	14.36	28.04	6.08
Mississippi	0.55	0.33	-4.64
Tennessee	15.08	15.65	0.34
East South Central Total	45.59	66.59	3.44
GRAND TOTAL	271.73	456.46	4.72

[1]Annual average growth rate, calculated on a basis of continuous compounding.

Table 4-2

BASE FORECAST OF 1980 SHIPMENTS FROM APPALACHIAN
PRODUCING DISTRICTS TO ELECTRIC UTILITIES,
ASSUMING NO MARKETS LOST TO RESIDUAL FUEL OIL

(Millions of Tons)

District	1969 Estimated Distribution	1980 Forecast Distribution	Average Annual Growth Rate (Percent)
1	31.67	48.25	3.83
2	7.84	11.64	3.59
3 + 6	30.92	55.16	5.26
4	37.04	67.71	5.48
7	1.47	1.48	0.01
8	55.61	82.83	3.62
13	11.19	16.00	3.25
TOTAL	175.74	283.07	4.33

Table 4-3

ANNUAL GROWTH RATES OF POWER GENERATION,
AND SHARE OF APPPALACHIAN COAL IN TOTAL UTILITY COAL USE,
BY STATE, 1970-1980

State, by Region	Average Annual Growth Rate of Electric Generation 1970-1980[1]	Share of App. Coal in Total Coal Use[2]
Connecticut	6.43	100.0
Massachusetts	6.43	100.0
New Hampshire	6.43	100.0
New England Total	6.43	100.0
New Jersey	6.43	100.0
New York	6.43	100.0
Pennsylvania	6.43	100.0.
Middle Atlantic Total	6.43	100.0
Illinois	9.29	*
Indiana	8.18	17.2
Michigan	8.18	93.7
Ohio	8.18	92.8
Wisconsin	9.29	7.5
East North Central Total	8.57	44.5
Delaware	6.43	100.0
District of Columbia	6.35	100.0
Florida	6.35	50.7
Georgia	6.35	50.7
Maryland	6.43	100.0
North Carolina	6.35	100.0
South Carolina	6.35	100.0
Virginia	6.35	100.0
West Virginia	8.18	100.0
South Atlantic Total	6.69	90.6
Alabama	6.35	63.1
Kentucky	8.18	14.1
Mississippi	5.84	63.1
Tennessee	6.35	47.6
East South Central Total	6.86	42.5
GRAND TOTAL	7.33	64.7

* = Less than 0.05 percent.

(Table continued on following page.)

82

Table 4-3 (Continued)

ANNUAL GROWTH RATES OF POWER GENERATION,
AND SHARE OF APPALACHIAN COAL IN TOTAL UTILITY COAL USE,
BY STATE, 1970-1980

SOURCES: [1]Electric Power Growth: Johnson, *op. cit.*, Chapter 10.

[2]Share of Appalachian coal estimated from *Bituminous
Coal and Lignite Distribution*, March 23, 1971.

coal consumption. With the slowdown in electricity growth (discussed further in Chapter 7), utilities in general are planning to add less capacity. Even by 1975, however, utilities in Virginia were planning almost 2000 megawatts of nuclear capacity by 1980, and the Tennessee Valley Authority was planning almost 3500 megawatts for Tennessee. Utilities in New York, Michigan, North Carolina and Alabama were planning to add nuclear capacity between 2000 and 3000 megawatts (per state).[1]

The U.S. Bureau of Mines publishes annual data on coal shipments from Bituminous Coal Producing Districts to states.[2] Data from this source for 1970 were used to distribute consumption by utilities in the different states among the Appalachian producing districts. The districts, defined by the Bituminous Coal Act of 1937, include all coal producing regions in the United States. The districts in Appalachia, dseigned in part to reflect homogeneous coal mining conditions, can be broadly summarized as follows:

1. Eastern Pennsylvania
2. Western Pennsylvania
3. Northern West Virginia
4. Ohio
6. Panhandle (West Virginia)
7. Southern Number 1 (parts of Virginia and southern West Virginia)
8. Southern Number 2 (parts of Virginia, West Virginia, eastern Kentucky, Tennessee and North Carolina)
13. Southeastern (all of Alabama and parts of Georgia and Tennessee).

Table 4-4 shows the fraction of coal received by electric utilities in each state from each Appalachian producing district in 1970. These shares were used to distribute the utility consumption in Table 4-1 among the producing districts in Table 4-2.

---

[1]Tabulated from National Coal Association, *Steam-Electric Plant Factors*, 1975 edition, Table 5.

[2]U.S. Bureau of Mines, *Mineral Industry Surveys*, "Bituminous Coal and Lignite Distribution, Calendar Year 1975".

Table 4-4

FRACTION OF COAL FROM APPALACHIAN PRODUCING DISTRICTS RECEIVED BY ELECTRIC UTILITIES IN EACH STATE, 1970

State	Districts						
	1	2	3 & 6	4	7	8	13
Connecticut	.985	0	.015	0	0	0	0
Massachusetts	.863	0	.038	0	0	.099	0
New Hampshire	.033	0	.967	0	0	0	0
New Jersey	.263	0	.615	0	.047	.075	0
New York	.407	.164	.371	.021	.011	.025	0
Pennsylvania	.538	.194	.268	0	0	0	0
Illinois	0	0	0	0	0	*	0
Indiana	0	0	0	.005	0	.012	0
Michigan	.005	.008	.092	.593	.001	.237	0
Ohio	.003	.013	.102	.694	0	.116	0
Wisconsin	.007	.004	.024	.028	0	.013	0
Delaware	.621	0	.374	0	0	.006	0
District of Columbia	.501	0	.009	0	.291	.200	0
Florida	0	0	0	0	0	.451	.056
Georgia	0	0	0	0	0	.451	.056
Maryland	.621	0	.374	0	0	.006	0
North Carolina	0	0	0	0	.008	.992	0
South Carolina	0	0	0	0	0	1.000	0
Virginia	.004	0	0	0	.096	.900	0
West Virginia	.115	0	.479	.091	0	.315	0
Alabama	0	0	0	0	0	0	.631
Kentucky	0	0	0	0	0	.141	0
Mississippi	0	0	0	0	0	0	.631
Tennessee	0	0	0	0	0	.455	.021

* = Less than .0005

Table 4-2 shows that, according to the extrapolated coal consumption, the coal-producing districts expected to increase their shipments of utility coal at the fastest rate are Districts 3, 4 and 6 (Ohio and northern West Virginia). Ohio ships mainly to the East North Central states. Districts 3 and 6 ship utility coal mainly to West Virginia itself, to the Middle Atlantic states, and to nearby states in the East North Central region. District 7 (Southern Number 1 -- Virginia and the southeastern part of West Virginia) ships very little utility coal, all of which goes to Virginia and the District of Columbia, whose combined coal consumption is expected to remain close to its 1969 level. Shipments from the other Appalachian producing districts are forecasted to grow at roughly the same rates -- between 3 and 4 percent per year.

Equilibrium Utility Demand for Coal

In this section, I assume that 1969 relative coal and oil prices are maintained. I first discuss the model and then use it to forecast coal demand by utilities on the assumption of no sulfur emission standards. I then discuss how the model is used to analyze the impact of different policies, focusing on sulfur-in-fuel restrictions.

Long-Run Equilibrium Demand

The model focuses on long-run utility demand. I use the term "long-run" to mean a period long enough for certain adjustments to occur. In particular, the time is long enough to construct a pipeline to carry residual fuel oil or to reconvert an oil-burning furnace to a coal-burning one. The period is substantially less than the 9 or 10 years required to plan, construct and put into operation a nuclear power plant.

86

"Long-run" is used here, additionally, to characterize fuel buyer'
expectations about the changes in relative prices. It is assumed, for
example, that, after an increase in the coal price of $0.02 per million
BTU (2¢/M BTU), the new delivered price of coal relative to that of residual
fuel oil is expected to persist for a long time. No attempt is made to
analyze here the small shifts in relative fuel use in response to
temporary price changes (although such shifts do occur, of course). Instead,
the analysis focuses on the major shifts in fuel use that may occur when
relative fuel prices change dramatically and persistently from their
traditional patterns.

The Approach in General

The model used to estimate the long-run demand curve is non-statistical.
Although it would be feasible to use available data to estimate a demand
curve by conventional econometric methods, such estimates would not be
very useful for the purposes of this study. The difficulty is that
historical experience does not encompass changes in relative coal prices
of the size that sulfur restrictions will bring about. For example, no
plants away from the seaboard have ever burned substantial amounts of
residual fuel oil. In New Hampshire, some small inland plants burn small
amounts of oil, and in other states some oil is typically burned for
peaking purposes. No observed price changes, however, have induced them
to switch to fuel oil for baseload purposes. Long-run demand coefficients
estimated from such data would provide a check, but there is no compelling

87

reason to expect them to be a more valid measure of the response to a
drastic increase in the relative cost of burning coal.[1]

The estimates of the long-run elasticity of demand for coal by
electric utilities (in states using Appalachian coal for this purpose)
are constructed on two basic assumptions. The first, and more critical,
assumption is that a steam-electric plant chooses, in the long run, to
burn coal only if it has the lowest average as-burned cost per BTU. This
assumption is strictly true only if the marginal cost of burning coal
is equal to its average cost over the relevant range of levels of fuel
consumption.

There are two sorts of conditions which could cause the marginal
to diverge from the average costs of burning coal. First, if the electric
utility is not a price-taker in the coal market, its purchases may bid up
the price it pays for coal. This possibility seems unlikely in most
situations, because there are so many purchasers of coal. Further, in the
long run, the supply curve of high-sulfur coal, even in local markets,
appears to be virtually horizontal, so that the utility's purchases
cannot affect the price.

Second, the average price paid by a utility is, as a rule, a mixture
of spot prices and prices under contracts of various lengths. If a utility
decided to reduce its coal use sharply, it would probably stop purchasing
in the spot market. There is, however, no reason to expect spot prices

---

[1]Although econometric estimates would provide a check on the estimates
derived from this model, there is no way of knowing whether they would be
too high or too low in the region outside the limits of observed price-
quantity relationships. If the econometrically estimated curve could be
validly extrapolated, these estimates would apply. There is no way of
knowing whether such a demand curve can be validly extrapolated beyond
the sample bounds, however.

to diverge systematically from the contract prices: on balance, therefore, this behavior would not cause the marginal price to differ from the reported average price. This relationship is most likely to hold when coal prices have been in rough equilibrium for some period. During transition periods, spot prices may be systematically above or below contract prices, since contract prices reflect past prices but also expected future prices.

The assumption that a utility will stop using coal altogether if its average as-burned cost is above that of residual fuel oil may be supported by another line of reasoning. The long-run demand for coal by an individual utility is quite likely to have a discontinuity at the point where the price of coal, in terms of equivalent heat content, is equal to the price of residual fuel oil. The source of this discontinuity is that coal cannot be burned very efficiently for peaking purposes, particularly in large units, because of the relatively high costs of starting up and shutting down a coal-burning unit.

For this reason, plants depending primarily on coal often burn some natural gas or fuel oil to meet peak demands, but few plants burning oil for base load generation burn substantial amounts of coal. Hence, the assumption was made that, if the average as-burned cost per BTU of coal exceeds that of residual fuel oil, the long-run response is to cut coal consumption to zero. If, on the other hand, the as-burned cost of residual fuel oil exceeds that of coal, then a coal-using plant is assumed to continue its use of coal (although it may also burn some fuel oil or natural gas).

There are, as noted earlier in this chapter, two parts of the sample of electric utility plants burning coal. First, for the states in which the electric utilities burn some coal, there are the 305 plants listed in the 1970 edition of *Steam -Electric Plant Factors* as burning 50,000 tons of coal a year or more. These plants account for over 99 percent of total coal used by electric utilities in those states. The cost of burning coal at each plant was taken to be the as-burned cost (in cents per million BTU) as reported in the same source. Use of this figure implicitly takes into account the minehead price of the coal, transportation costs, and heat content.

The second part of the sample consists of all the new coal-burning plants expected to be in operation by 1973. Because these plants have not yet had any operating experience, the 1969 coal cost had to be estimated for them. When the location of the new plant was known, this cost was taken to be the cost at the nearest electric plant operating in 1969. When the location was not known, the cost was taken as the average for all the plants of the electric power company planning the plant. An electric company, in its actual decision as to the location of a new plant, would have a good idea of the costs of burning coal, but the average will have to serve for our purposes.

The second assumption concerns the delivered price of imported residual fuel oil. Because large quantities of imported residual fuel oil have been consumed in only a few states on the eastern seaboard, it was necessary to estimate prices in ports of entry (and to assign such ports to inland electric plants) and transportation costs from the port to inland generating stations.

First, ports were selected on the basis of their having facilities to handle residual fuel oil. The list of these ports appears in Table 4-5. A port was included if a price for residual fuel oil was shown for its terminal in *Platt's Oil Price Handbook and Oilmanac* during 1969. Most of these ports do not, at present, have facilities to unload the supertankers that could bring in large quantities of oil at minimum per-barrel transport costs. Utilities dependent on such ports might face difficult obstacles, in the short run, in buying oil at the hypothesized price if they converted their coal burners. Such difficulties are not explicitly considered in the analysis here, because they could be resolved in a few years. For example, offshore unloading facilities could be constructed if it were clear to the utilities that they would find residual fuel oil the cheapest fuel over a long time period. As such construction, including planning, might take three to four years, it could easily be done by 1980.

The as-burned costs for residual fuel oil of electric utilities in the port of entry (if available) or the average of the state in which the port is located were taken from *Steam-Electric Plant Factors,* 1970 edition. Use of these costs implicitly takes into consideration the costs of delivery from the terminal to a utility's burners. The base prices of residual fuel oil, as-burned, assumed for utilities in each state, are shown in Table 4-6. For inland utilities, transportation costs from the port to the plant were estimated and added to this average cost.

91

Table 4-5

EAST AND GULF COAST PORTS
FOR RESIDUAL FUEL OIL

Baltimore, Md.

Bangor, Me.

Baton Rouge, La.

Boston, Ma.

Charleston, S.C.

Houston, Tex.

Jacksonville, Fla.

Miami, Fla.

New Haven, Conn.

New Orleans, La.

New York, N.Y.

Newark, N.J.

Norfolk, Va.

Philadelphia, Pa.

Portland, Me.

Providence, R.I.

SOURCE: *Platt's Oil Price Handbook and Oilmanac,* as cited in U.S.
          Bureau of Mines, *Oil Availability by Sulfur Levels, op. cit.,*
          Table 19.

Table 4-6

ASSUMED PRICES FOR RESIDUAL FUEL OIL,
AS BURNED COSTS, PORT OF ENTRY
(Cents per million BTU)

State, by Region	Base Level Price for Fuel Oil[1]
New England	
Connecticut	28.3*
Massachusetts	28.2*
New Hampshire	28.0*
Middle Atlantic	
New Jersey	34.5*
New York	40.0**
Pennsylvania	30.0**
East North Central	
Illinois	30.1
Indiana	30.1
Michigan	30.1
Ohio	30.3
Wisconsin	30.1
South Atlantic	
Delaware	31.8*
District of Columbia	41.2*
Florida	30.1*
Georgia	34.4*
Maryland	30.1*
North Carolina	33.0
South Carolina	33.0
Virginia	28.4*
West Virginia	28.4
East South Central	
Alabama	30.1
Kentucky	28.4
Mississippi	30.1
Tennessee	28.4

Notes on following page.

Table 4-6 (Continued)

ASSUMED PRICES FOR RESIDUAL FUEL OIL,
AS BURNED COSTS, PORT OF ENTRY

(Cents per million BTU)

*Actual average cost per BTU, as burned, in that state in 1969.

**Actual average cost per BTU, as burned, in metropolitan area, 1969.

Values not marked with * or ** were taken as values in estimated port state.

SOURCE: *Steam-Electric Plant Factors*, 1970 edition, Table 2. Value shown is either state average (metropolitan area, if given) or average in state of port of entry.

To estimate transportation costs, estimates were made of the shortest distance by pipeline, barge, or combination, from the nearest port to each steam-electric plant. Because there has been no experience with transporting residual oil inland in large quantities, two assumptions were made as a probable range for such costs. These costs are discussed in detail in the previous chapter, so they are simply listed here:

	Low Cost	High Cost
Pipeline:	.016¢/M BTU per mile (1 mill per barrel mile)	.033¢/M BTU per mile (2 mills per barrel mile)
Barge:	.016¢/M BTU per mile (1 mill per barrel mile)	.167 ¢/M BTU per mile (10 mills per barrel mile)

In the scenarios analyzed in Chapter 6, these transportation costs were combined with two assumptions about residual fuel oil prices. One set of scenarios, therefore, assumed low transport costs and 1969 residual prices.[1] This case implies, other things equal, the least growth in coal use by electric utilities. The second set of scenarios assumed high transport costs and residual fuel oil prices 4¢/M BTU above their 1969 level. Because of the substitutability of European-refined oil,

---

[1]For convenience's sake, all of the costs are taken to be in 1969 dollars. The implicit assumption is that inflation does not change relative costs of different fuels or of different modes of transportation. Imported oil prices are also, of course, sensitive to changes in the exchange rate, but such changes will probably be small, and, in any case, prediction of such changes is beyond the scope of this study. Consequently, it is simply assumed that the world price of the U.S. dollar does not change.

this amount represents a ceiling on unilateral increases in Venezuelan oil prices.[1] It is not, of course, a ceiling if OPEC as a whole raises oil prices. This latter case is discussed in Chapter 7.

A number of uncertainties surround the question of expanded use of residual fuel oil by electric utilities. First, uncertainty about the tax policy of Venezuela causes uncertainty about the future price of residual fuel oil. We have tried to bound this uncertainty with the 1969 price on one side and the 1969 price plus 4¢/M BTU on the other. Second, because of the high initial outlays required for construction of a residual fuel pipeline or of an off-shore unloading facility, these investments would be made only if the price of residual fuel oil was expected to be below that of coal for a long period of time. To the extent that the high transport costs overstate the actual transport cost, the difference might be taken as an estimate of reluctance to construct such pipelines or facilities.

Third, because the costs to utilities of being without fuel are so high, electric utilities often are willing to pay some price to maintain alternative sources of fuel supply. To the extent that foreign oil supplies (even from the Western Hemisphere) are viewed as unstable or risky, the cost advantage of residual oil must be greater to induce coal-burning utilities to switch to oil. This differential could be built into the model, but it was not, as no estimate of its importance was available.

---

[1]In the short run, electric utilities may be reluctant to change suppliers, as there are transactions and market information costs to doing so. Over a period as long as seven or eight years, however, those costs are very small relative to the potential gains from lower prices, so it is reasonable to assume that, if Venezuelan prices rose above the transport differential (assuming similar prices at the refinery), electric utilities would shift to European sources.

The prices of residual fuel oil shown in Table 4-6, with the possible exception of those for New York, are for oil without any sulfur guarantee. These prices are those that would be paid if no sulfur emission standard had to be met, and they are the ones used in the scenarios analyzing the status quo and the tax of $0.25 per ton on coal.

Desulfurization of residual fuel oil, however, incurs additional costs, and these costs need to be considered in analyzing a utility's fuel choice under a 2.0 percent and a 0.7 percent sulfur-in-fuel standard. It is assumed that, to meet a 2.0 percent standard, residual fuel oil can be bought without incurring any additional costs. There are large amounts of Venezuelan residual fuel oil containing less than 2.5 percent sulfur, so that essentially no increase in price (or only a very trivial one) would be needed to meet whatever amount of oil at that sulfur level would be demanded.[1]

To meet a 0.7 percent standard, the oil could be desulfurized at the refinery. The methods used to desulfurize residual fuel oil below 1.0 percent have not, thus far, been successful on high-sulfur Venezuelan crudes, whose high metal content tends to poison the catalysts used. The costs of desulfurizing residual oil to this level are not well known, then, but it seems quite probable that, by 1980, the costs would be in the range of $0.50 to $1.00 per barrel. Table 4-7 summarizes recently published information on desulfurization costs; the cost used in the scenarios -- $0.75 per barrel -- appears to lie in the range of estimates shown.

---

[1]This is not to say that the supply of Venezuelan residual oil is highly elastic, but only that a 2.0 standard imposes no additional constraint on supply.

Table 4-7

COSTS OF DESULFURIZING RESIDUAL FUEL OIL

Date	Source	Type of Process	Type of Fuel	Sulfur Level		Cost/bbl (Dollars)	Each 1% Removed
				From	To		
4/69	Oil & Gas Journal	Direct	Middle Eastern	≈3.5	1.10		.20
7/69	Oil & Gas Journal	Indirect	Venezuelan	2.5	1.00	.30	
2/70	Oil & Gas Journal	Indirect	Venezuelan	2.5	1.00	.30	
5/71	Oil & Gas Journal	Direct	Middle Eastern	"high"	.40		.21
7/71	Crentz & Deurbrouck			2.5	.50	1.14	
				2.5	.50	1.71	

SOURCES: *Oil and Gas Journal*, April 21, 1969, p. 54.
July 14, 1969, p. 31.
February 9, 1970, p. 25.
May 17, 1971, p. 85.

William L. Crentz and Albert W. Deurbrouck, "Coal and the Sulfur Problem", Eighth World Energy Conference (Bucharest: July 1971), p. 2.

The Model

In symbols, the model can be set forth as follows:

Let $C(J)$ = 1969 as-burned cost of coal in cents per million BTU at electric utility plant J;

$E(J)$ = 1969 as-burned cost of residual fuel oil in the assumed port of entry for plant J, or actual as-burned costs of other utilities in the same city (if available), in cents per million BTU;

$P(J)$ = miles of pipeline from the port to electric plant J;

$B(J)$ = barge miles from the port to electric plant J;

$P(J) + B(J)$ = total distance oil travels from port to plant J;

$R(J) = E(J) + [A1.P(J) + A2.B(J)]$ = 1969 as-burned cost of residual fuel oil at plant J, where $A1$ = cost per mile by pipeline; $A2$ = cost per mile by barge;

$Q(J,T)$ = amount of coal consumed at plant J in year T, in thousands of tons, if coal is the fuel chosen; and

$W(J,T)$ = predicted coal use at plant J in year T, in thousands of tons.

The relationship that forms the basis for all of the scenarios is:

$$W(J,T) = \begin{cases} Q(J,T) \; if \; C(J) \overset{\leq}{=} R(J) \\ 0, \; if \; R(J) < C(J). \end{cases} \tag{4}$$

If there are N_K electric utility plants in state K, total coal consumption in that state in year T can be written as:

$$Z(K,T) = \overset{N_K}{\underset{J}{\Sigma}} W(J,T). \tag{5}$$

99

The coal consumption by electric utility plants in state K in year T is then allocated among the Appalachian producing districts according to 1970 shipments patterns:

$$X(D,K,T) = \frac{X(D,K,1970)}{Z(K,1970)} \; Z(K,T), \qquad\qquad (6)$$

where $X(D,K,T)$ is shipments to state K from district D in year T. Table 4-4, discussed above, shows the share of each district's shipments to each state in which electric utilities burn some Appalachian coal. That is, the entries in Table 4-4 are the values of $X(D,K,1970)/Z(K,1970)$ for each state K and each district D. With different sulfur standards, or with policies that affect differentially the costs of strip and underground mining, for instance, these patterns would change, but I have not attempted here to adjust for such changes.

Total shipments from district D to electric utilities, for the 24 states in which utilities burn some Appalachian coal, can be written as:

$$X(D,T) = \sum_{K=1}^{24} X(D,K,T) \qquad\qquad (7)$$

Closing the Model

The model as described above can be used to estimate the amount of coal that would be demanded at different relative prices of coal and fuel oil. A supply function for coal is also needed, however, to forecast the equilibrium level of consumption. In this chapter and Chapter 6, I assume that the long-run supply function for coal is horizontal at its 1969 real price, plus $0.75 per ton to adjust for the costs imposed by the 1969 Federal Coal Mine Health and Safety Act (FCMHSA).

For reasons that have been discussed in detail elsewhere, it is very difficult to estimate an econometric supply function for coal.[1] Other sorts of data, notably the large number of coal deposits whose locations are well known, suggest that in the long run, output can either expand or contract (over the expected range) without any change in the real price of coal. In this context, "long run" is a period of about three years, the average length of time required to open a new large underground mine. The large number of coal mining companies, and the ease of entry into and exit from the industry, suggest that market power poses no obstacle to expansion of coal supply.[2]

A Check of the Model

For each utility, the cost of coal was compared with the cost of residual fuel oil, under a variety of assumptions about the increase in coal prices over their 1969 level. Coal use was estimated for each utility according to the procedure described above, and coal use was aggregated for consuming states and regions.

---

[1]See, for example, Charles River Associates, Incorporated, *The Economic Impact of Public Policy on the Appalachian Coal Industry and the Regional Economy* (Cambridge: January 1973), Volume I, "Profile of the Appalachian Coal Industry and Its Competitive Fuels", pp. 247-251.

[2]A good survey of recent studies of coal supply is Richard L. Gordon, *Economic Analysis of Coal Supply: An Assessment of Existing Studies* (Palo Alto, CA: Electric Power Research Institute, May 1975). Many of the studies reviewed focus on the supply of low sulfur Appalachian coal, while the assertions made here refer to the supply of *high* sulfur Appalachian coal. Over the output ranges implied by the analysis in this study, the assumption seems plausible.

One check on the estimates made this way is a comparison of the 1969 actual coal consumption with the estimate of coal consumption when the price increase is assumed to be zero. This comparison appears in Table 4-8. This check, although it is useful in evaluating the validity of the model's assumptions for past behavior, does not imply accuracy for forecasting purposes.

The estimates are exactly the same as the actual in the East South Central region and almost exactly the same in the East North Central region. The major differences occur for New England, New Jersey, Pennsylvania, Maryland and Virginia. The total estimated is about 10 percent less than the actual. Thus the model appears to do fairly well, especially since some of the states for which it overpredicts were moving away from coal.[1]

To illustrate how an increase in the price of coal would induce electric utilities to shift to residual fuel oil, in the long run, estimates are presented of what coal consumption by electric utilities would have been in 1969, if they had been in equilibrium and if the price of coal had included the long-run costs of meeting the safety requirements of the Federal Coal Mine Health and Safety Act of 1969.

---

[1] By 1973, for example, coal use in Connecticut and Massachusetts had fallen almost to zero, while coal use in New Jersey, Delaware, The District of Columbia, Maryland, and Virginia had fallen by about 50 percent. (*Steam-Electric Plant Factors*, 1974 edition, Table 2).

102

Table 4-8

COMPARISON OF 1969 CONSUMPTION AND ESTIMATED LONG-RUN EQUILIBRIUM CONSUMPTION FOR 1969, AT 1969 RELATIVE PRICES AND GENERATION LEVELS
(Millions of Tons)

State, by Region	Actual 1969 Coal Use[1]	Estimated Long-Run Level (Low Oil Transport Costs)
Connecticut	2.11	0.00
Massachusetts	1.88	0.00
New Hampshire	0.93	0.00
New England Total	4.93	0.00
New Jersey	4.29	1.64
New York	12.85	12.85
Pennsylvania	26.31	19.89
Middle Atlantic Total	43.44	34.38
Illinois	29.53	29.53
Indiana	22.53	22.53
Michigan	19.89	19.88
Ohio	33.53	33.53
Wisconsin	9.24	9.14
East North Central Total	114.72	114.62
Delaware	1.34	0.95
District of Columbia	0.78	0.38
Florida	4.53	4.05
Georgia	7.49	7.49
Maryland	6.76	1.29
North Carolina	16.21	15.18
South Carolina	3.50	3.43
Virginia	7.89	3.06
West Virginia	14.55	14.00
South Atlantic Total	63.05	49.83
Alabama	15.60	15.60
Kentucky	14.36	14.36
Mississippi	0.55	0.55
Tennessee	15.08	15.08
East South Central Total	45.59	45.59
GRAND TOTAL	271.73	244.42
Estimated Total as Percent of Actual Total	99.30	89.40

Footnote on following page.

103

Table 4-8 (Continued)

COMPARISON OF 1969 CONSUMPTION AND ESTIMATED LONG-RUN
EQUILIBRIUM CONSUMPTION FOR 1969, AT 1969
RELATIVE PRICES AND GENERATION LEVELS

[1]Entries may differ slightly from values given in Chapter 3 because of omission of plants burning less than 0.05 million tons of coal in 1969.

These costs were not, of course, incurred until after 1969 and were not reflected in the 1969 as-burned costs reported by electric utilities. This illustration is therefore somewhat hypothetical, and it is intended to demonstrate how the model works, free of the complexities introduced by forecasts of net electricity generation.

The costs of meeting the safety requirements are expected to add about $0.75 per ton to the price of coal.[1] On the assumption that coal contains about 25 million BTU's per ton, this cost is equivalent to about 3¢/M BTU.[2] Table 4-9 presents the estimates of the amount of coal that would have been demanded by electric utilities in 1969, in equilibrium, if these costs had been included.

Table 4-9 shows that, for a relatively small increase in the price of coal, there is not much difference between the estimates using low residual price and transport costs and those using high residual price and transport costs, except for a difference of about 6 million tons of consumption in the South Atlantic region.

A comparison of Table 4-8 with Table 4-9, however, shows, for the same assumptions about residual oil costs, that an increase in the coal price of 3¢/M BTU leads to a decrease in coal consumption of about

---

[1]Charles River Associates, Incorporated, Volume I, pp. 336-345. The actual costs of compliance with FCMHSA appear to be substantially higher, even in real terms.

[2]Appalachian utility coal in 1969 averaged between 23 million and 26 million BTU per ton, but, given the level of accuracy due to available data, no real error is introduced by the convenient assumption of 25 million, which permits easy conversion from dollars per ton to cents per million BTU: $1.00 per ton = 4¢/M BTU.

Table 4-9

ESTIMATE OF 1969 COAL CONSUMPTION BY ELECTRIC
UTILITIES, GIVEN AN INCREASE IN THE COAL PRICE OF 3¢/M BTU
(Millions of Tons)

State, by Region	Low Residual Fuel Oil Transport Cost and Price Assumptions	High Residual Fuel Oil Transport Cost and Price Assumptions
Connecticut	0.00	0.00
Massachusetts	0.00	0.00
New Hampshire	0.00	0.00
New England Total	0.00	0.00
New Jersey	0.00	0.13
New York	8.02	9.12
Pennsylvania	17.60	17.60
Middle Atlantic Total	25.61	26.84
Illinois	29.53	29.53
Indiana	22.53	22.53
Michigan	19.88	19.88
Ohio	33.23	33.52
Wisconsin	9.15	9.24
East North Central Total	114.32	114.72
Delaware	0.82	0.87
District of Columbia	0.33	0.33
Florida	0.00	0.00
Georgia	7.08	7.08
Maryland	0.48	1.37
North Carolina	14.05	15.24
South Carolina	2.45	3.42
Virginia	2.99	6.29
West Virginia	14.55	14.55
South Atlantic Total	42.75	49.16
Alabama	15.60	15.60
Kentucky	14.36	14.36
Mississippi	0.55	0.55
Tennessee	15.08	15.08
East South Central Total	45.58	45.58
GRAND TOTAL	228.27	236.30

16 million tons, with virtually all of the decline occurring in the Middle Atlantic and South Atlantic states.

The model suggests that the transportation costs of residual oil prevent it from making inroads on coal's market except along the eastern seaboard. Figures in Table 4-8 suggest, however, that, with relative fuel prices at their 1969 level, there are a number of plants along the eastern seaboard that would find it profitable to convert to residual fuel oil.

The hypothetical decreases shown in Table 4-9 would affect different Appalachian producing districts quite differently. When the 1970 pattern of shipments is applied to actual 1969 coal use (Table 4-8) and to the totals shown in Table 4-9, the resulting distribution is as shown in Table 4-10.

As might be expected, the districts most severely affected (at least in their shipments of utility coal) are those closest to the eastern seaboard -- Districts 1 and 2 in Pennsylvania, and Districts 3 and 6 in West Virginia. District 7, the easternmost West Virginia district, is, in percentage terms, also severely affected, but it ships such small quantities of utility coal that the absolute drop is less than 1 million tons.

Forecast for 1980, Assuming Equilibrium at 1969 Relative Prices

The forecast discussed in this section provides a measure of electric utility consumption of Appalachian coal at 1969 relative prices in the absence of sulfur emission restrictions. It is thus a yardstick by which the impact of these restrictions can be measured, rather than a prediction.

107

Table 4-10

EFFECT OF 3¢/M BTU PRICE INCREASE ON SHIPMENTS FROM APPALACHIAN
PRODUCING DISTRICTS TO ELECTRIC UTILITIES, 1969 COAL USE LEVELS

(Millions of Tons)

District	(1) 1969 Shipments to Electric Utilities	(2) Estimated Distribution of Actual 1969 Coal Use	(3) Estimate From Table 4-9, Low Residual Oil Costs	(3)/(2) (Percent)	(5) Estimate From Table 4-9, High Residual Oil Costs	(5)/(2) (Percent)
1	30.72	31.67	15.68	49.5	16.77	53.0
2	9.15	7.84	5.36	68.4	5.54	70.7
3 & 6	35.67	30.92	20.58	66.6	21.45	69.4
4	37.41	37.04	36.72	99.1	36.95	99.8
7	1.45	1.47	0.60	40.8	0.94	63.9
8	60.39	55.61	44.98	80.9	50.17	90.2
13	11.07	11.19	10.91	97.5	10.91	97.5
TOTAL OF DISTRICTS SHOWN	185.86	175.74	134.83	76.7	142.73	81.2

NOTES: Estimates were computed from state consumption in Tables 4-8 and 4-9, distributed according to the 1970 shipments pattern.

The source for column (1), shown here only for purposes of comparison (since it is a different sample from the one used to compute coal used by electric utilities), is from U.S. Bureau of Mines, *Minerals Yearbook*, 1969 edition, p. 356.

Because the safety requirements of the Federal Coal Mine Health and Safety Act (FCMHSA) have already taken effect, the base forecast for 1980 was calculated on the assumption that the real price of coal will be 3¢/M BTU higher than its 1969 level. The comparison of fuel cost at each plant -- actual or planned -- was computed, and its 1980 choice of fuel was determined in the manner discussed above. If coal was found to be the least-cost fuel, then the coal needed for the forecasted electric generation was added to its state coal-use total. If residual fuel oil was found to be the cheapest fuel source, then no coal was added. The forecasts of net electric generation and coal use per kilowatt hour were discussed in the first part of this chapter.

In effect, then, the base forecasts are a combination of the straightforward projections of the first section of this chapter and the demand model, with an allowance made for increased coal prices from the 1969 safety law.

These forecasts are shown, for both low and high residual transport costs, in Table 4-11.[1] A comparison with Table 4-1, which assumes that no markets are lost to residual fuel oil, shows that all of the decreases in coal consumption occur along the Atlantic seaboard. Utilities in the Midwest and East South Central states, on the other hand, consume virtually the same amounts under all three sets of forecasts.

---

[1]The assumption of low transport costs also assumes 1969 residual fuel oil prices, while the high cost assumption assumes an increase in residual fuel oil prices of 4¢/M BTU above their 1969 level. For expository convenience, however, the assumptions will be referred to as "low cost" and "high cost".

Table 4-11

BASE FORECASTS FOR 1980, ASSUMING COAL PRICE
INCREASE OF 3¢/M BTU DUE TO MINE SAFETY
REQUIREMENTS

(Millions of Tons)

State, by Region	Low Residual Oil Costs	High Residual Oil Costs
Connecticut	0.00	0.93
Massachusetts	0.00	0.00
New Hampshire	0.00	0.00
New England Total	0.00	0.93
New Jersey	0.00	3.00
New York	9.18	14.73
Pennsylvania	33.01	35.98
Middle Atlantic Total	42.20	53.71
Illinois	60.46	60.46
Indiana	42.12	42.12
Michigan	35.31	35.31
Ohio	60.18	60.58
Wisconsin	14.63	14.78
East North Central Total	212.71	213.26
Delaware	1.19	2.24
District of Columbia	0.48	0.48
Florida	1.02	7.62
Georgia	12.92	13.31
Maryland	0.37	8.18
North Carolina	18.08	20.24
South Carolina	3.48	4.08
Virginia	3.22	6.17
West Virginia	40.71	41.50
South Atlantic Total	81.48	103.83
Alabama	22.58	22.58
Kentucky	28.04	28.04
Mississippi	0.33	0.33
Tennessee	15.65	15.65
East South Central Total	66.59	66.59
GRAND TOTAL	402.98	438.32

The other regions are forecasted to lose some coal relative to the naive, noncompetitive projection. For example, coal used by New England electric utilities is forecasted to fall virtually to zero under either assumption about residual fuel oil transport costs. This loss amounts only to about 6 million tons relative to the noncompetitive projection.

Consumption in the Middle and South Atlantic states, on the other hand, is relatively sensitive to the assumption about residual oil costs. Under the high cost assumption, coal use in the Middle Atlantic states falls by about 7 million tons, but under the low cost assumption the loss is about 18 million tons, mainly in New York and Pennsylvania. In the South Atlantic states, about 7 million tons are lost under the high cost assumption, while under the high cost assumption 29 million tons are displaced by residual fuel oil. The distribution of consumption among Appalachian producing districts for the three cases is shown in Table 4-12.

As in Table 4-10, production declines most in those districts shipping coal towards the eastern seaboard -- Districts 1 and 2 in Pennsylvania and District 7 in West Virginia. Under the assumption of high residual oil costs, however, less than 10 percent of extrapolated electric utility shipments are lost in the base forecast. If low residual costs more nearly reflect these costs as seen by utilities, however, nearly 20 percent of Appalachian production for electric utilities is displaced by residual fuel oil.

Table 4-12

DISTRIBUTION OF SHIPMENTS BY APPALACHIAN PRODUCING DISTRICTS
TO ELECTRIC UTILITIES, 1980 FORECASTS

(Millions of Tons)

District	(1) No Markets Lost to Residual Fuel Oil	(2) Estimate From Table 4-11, Low Residual Oil Costs	(2)/(1) (Percent)	(5) Estimate From Table 4-11, High Residual Oil Costs	(5)/(1) (Percent)
1	48.25	27.92	57.9	39.09	81.0
2	11.64	9.03	77.6	10.52	90.4
3 & 6	55.16	42.06	76.3	50.50	91.6
4	67.71	67.24	99.3	67.71	100.0
7	1.48	0.72	48.7	1.22	82.4
8	82.83	70.90	85.6	80.16	96.8
13	16.00	15.58	97.4	15.97	99.8
TOTAL OF DISTRICTS SHOWN	283.07	233.45	82.5	265.18	93.7

SOURCE: State consumption forecasts of Table 4-11 and Table 4-1, distributed among Appalachian producing districts according to 1970 pattern of shipments.

The forecasts in Table 4-11 suggest that, at 1969 relative prices, even in the absence of other policies affecting the coal industry, some electric utility plants along the eastern seaboard would have converted to residual fuel oil. Table 4-11 estimates how these conversions affect production of coal for the electric utility market in Appalachia.

Tables 4-13 and 4-14 show, at various coal prices above 1969 levels, how much Appalachian coal would be demanded in 1980, at low and high residual fuel oil transportation costs, respectively. The elasticity estimates shown are arc elasticities, measured between successive price increases, and calculated as

$$E = D \ (\log Q)/D \ (\log P), \tag{8}$$

where E is elasticity, D stands for the change in a variable, Q refers to quantity demanded, and P to the price. The base price was 28.0¢/M BTU, the average coal price paid by electric utilities burning Appalachian coal. Total demand in these tables is simply the sum of electric utility demand plus other demand. Elasticity is computed on the basis of total demand. These relationships are graphed in Figures 4-1 and 4-2.

Changes in the Model for Other Scenarios

This section discusses only changes in the framework of the model for the analysis of different policies; the parameters assumed and the results are presented in detail in Chapter 6. Only minor changes in the model are needed for scenarios in which the as-burned cost of all coal changes by the same amount. If the cost increases by 3¢/M BTU due to mine safety regulations, then in the earlier notation, if

113

Table 4-13

FORECASTS OF 1980 DEMAND FOR APPALACHIAN COAL,
AT VARIOUS PRICES ABOVE 1969 PRICE LEVELS

(Low Residual Fuel Oil Price and Transport Costs)

¢/M BTU Above 1969 Level	Electric Utility Demand for Appalachian Coal	Total Demand for Appalachian Coal	Elasticity
0	249.00	487.22	--
3	233.45	471.68	-0.32
5	220.15	458.37	-0.46
7	184.34	422.57	-1.38
9	167.39	405.61	-0.74
11	141.31	379.53	-1.26
13	85.63	323.86	-3.17
15	48.19	286.41	-2.58
17	37.35	275.58	-0.85
19	31.66	269.88	-0.48
21	21.82	260.04	-0.89
23	4.32	242.55	-1.74

114

Figure 4—1

ESTIMATED DEMAND CURVE FOR COAL
(Low residual fuel oil price and transport cost assumptions, 1980 fossil generation levels)

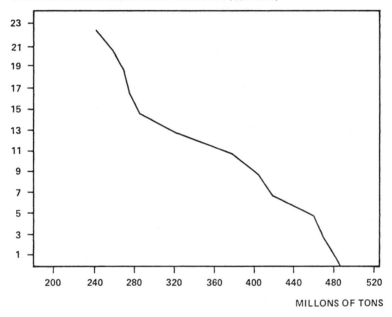

INCREASE IN COAL COSTS OVER 1969 LEVEL (¢ /M BTU)

MILLONS OF TONS

Table 4-14

FORECASTS OF 1980 DEMAND FOR APPALACHIAN COAL,
AT VARIOUS PRICES ABOVE 1969 PRICE LEVELS

(High Residual Fuel Oil Price and Transport Costs)

¢/M BTU Above 1969 Level	Electric Utility Demand for Appalachian Coal	Total Demand for Appalachian Coal	Elasticity
0	272.22	512.46	--
3	265.18	503.40	-0.18
5	259.42	497.64	-0.18
7	248.33	486.55	-0.38
9	244.22	482.45	-0.15
11	236.84	475.06	-0.29
13	224.89	463.11	-0.51
15	206.93	445.16	-0.83
17	186.79	425.02	-1.02
19	168.14	406.36	-1.03
21	140.40	378.62	-1.70
23	118.86	357.08	-1.46

Figure 4—2

ESTIMATED DEMAND CURVE FOR COAL
(High residual fuel oil price and transport cost assumptions, 1980 fossil generation levels)

INCREASE IN COAL COSTS OVER 1969 LEVEL (¢/M BTU)

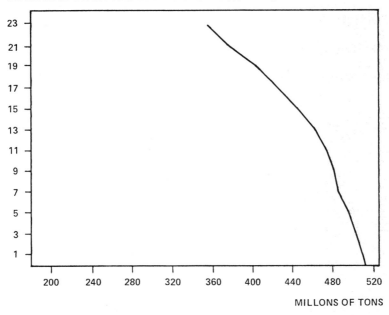

MILLONS OF TONS

117

$$C(J) + 3.0 \overset{\leq}{=} R(J), \tag{9}$$

the utility will burn $Q(J,1980)$ tons of coal. If $C(J) + 3.0 > R(J)$, the utility will burn oil instead at that plant. The other set of scenarios analyzed in this way is a 0.7 percent sulfur-in-fuel standard.

Analysis of a 2.0 percent sulfur-in-fuel standard requires somewhat more complex alterations to the model. The additional costs of burning Appalachian coal are assumed to be only $0.75 per ton (because it can meet the standard by being washed), while the extra costs for Midwest coal appear to be about $4.00 per ton (because a stack-gas device is required).[1] Both Appalachian and Midwest coal are burned by electric utilities in Illinois, Indiana, and Wisconsin, so that there is a possibility that, to meet a 2.0 percent standard, electric utilities in those states would buy Appalachian rather than Midwest coal.[2]

Not all electric utility plants in the Midwest can burn Appalachian coal, however, because with its high ash-fusion temperature, it forms into a solid mass in a wet-bottom furnace, requiring an expensive shut-down and clean-up process.[3] In 1969, only about 21 million tons of coal

---

[1]This cost is slightly less than the $5 per ton discussed above, because it assumes desulfurization only to a level equivalent to 2.0 per cent. As discussed in Chapter 7, these estimates understate substantially the desulfurization costs estimated as of the mid-1970's.

[2]In the other East North Central states, Ohio and Michigan, electric utilities already burn Appalachian coal almost exclusively.

[3]See, for example, U.S. Bureau of Mines, *Restrictions on the Uses of Coal* (Washington: 1971), p. 11. In a wet-bottom furnace, the coal ash becomes liquid and is removed in molten form from the combustion chamber. A low ash-fusion temperature is essential to this process.

were burned in wet-bottom furnaces in those three states.[1] At the time the research was done, no data were available on exactly which plants use wet-bottom furnaces, so that no direct modification could be made to the model of spatial competition used to analyze demand.[2]

In the results of the scenarios, use of midwestern coal in Indiana and Illinois was greater than 1969 wet-bottom furnace consumption.[3] Although, depending on the geographical distribution of wet-bottom furnaces, the procedure used (described below) may overstate Appalachian coal use in those states, it is as least possible that Appalachian coal could be burned in those states in the estimated amounts.

The model used was as follows. For utilities already burning coal only from Appalachian districts, the condition for continuing to burn coal is:

$$C(J) + 3.0 \leqq R(J),\tag{10}$$

where $C(J)$ (in cents per million BTU) already includes mine safety costs. As burning residual oil requires no additional outlays to meet a 2.0 percent sulfur standard, the 3¢/M BTU corresponds directly to $0.75 per ton coal washing costs.

---

[1]About 10 million tons in Illinois, 8 million tons in Indiana, and 3 million tons in Wisconsin. *Ibid.*, p. 8.

[2]Information has since been published on this question in Federal Power Commission, *Steam-Electric Plant Air and Water Quality Control Data for the year ended December 31, 1969, Summary Report* (Washington: February 1973).

[3]Appalachian coal use in Wisconsin is, therefore, probably overstated by about 3 million tons, but the error is negligible relative to total Appalachian production.

For utilities in Illinois, Indiana, and Wisconsin, however, the comparisons were somewhat more complex. A utility was assumed to burn Appalachian coal only if both of the following conditions were met:

$$A + T(J) + 3.0 \stackrel{\leq}{=} C(J) + 16.0$$

and (11)

$$A + T(J) + 3.0 \stackrel{\leq}{=} R(J).$$

In this notation, $C(J)$ and $R(J)$ are as above, 16.0 represents the additional cost (in cents per million BTU) of a stack-gas device to permit the burning of midwestern coal, A represents the as-burned costs at a utility in East Kentucky or western West Virginia (where transport costs from the minehead are negligible) and $T(J)$ represents the cost (in cents per million BTU) of transporting the coal by unit train from East Kentucky to the midwestern utility in question.

The as-burned costs of Appalachian coal close to the minehead were determined by looking at as-burned costs in 1969 of minemouth generating plants, and were taken as $0.203 per million BTU. Transport costs were determined by computing the rail mileage from East Kentucky to the utility plant, applying the rail cost model of Chapter 3, and converting the units into cents per million BTU. The values of other variables in that model -- average loading and unloading time, and share of coal at the utility -- were taken at their sample means.

120

Chapter 5

NON-UTILITY DEMAND FOR APPALACHIAN COAL

In this chapter I analyze and forecast the demand for Appalachian
coal by sectors other than the electric utility industry. The uses
include coking demand, export demand, retail demand (primarily home
heating), and a heterogeneous grouping called "other manufacturing and
mining".

I chose to model demand in all of these uses as insensitive to
relative fuel prices. This assumption is approximately correct for the
demand for coal for coking. Export demand is more sensitive to coal
prices, because of alternative foreign sources of coal supply (such
as Poland and Australia), but it is probably dominated by energy policies
of importing countries. Retail demand is such a small amount that including a price effect would not change the forecasts materially. Demand
for coal by other manufacturing and mining includes some uses that
are sensitive to relative fuel prices and others that are not. This
category accounts for about 15 percent of coal consumption, and I found
that demand in this sector could be satisfactorily explained without
reference to relative prices.

121

I turn now to a discussion of each of these uses.

Coking Demand for Coal

About 20 percent of U.S. coal consumption is in the production
of coke, most of which, in turn, is used in blast furnaces to make pig
iron. Because there are no direct substitutes for coke in this process,
the demand for coal for coking purposes follows very closely the
production of pig iron. Although relative fuel prices may lead in
the long run to the discovery and introduction of processes using less
coke, in the short run, changes in relative fuel prices do not substantially
affect the amount of coke required per ton of steel.

Coke costs represent a relatively small fraction of the price of
steel. On the basis of 1970 input ratios and 1976 prices, coke accounts
for about 18 percent of the price of steel.[1] Changes in blast furnace
practices have, however, led to a gradual reduction in coke requirements
over time. As a result, although pig iron production in 1970 was about
16 percent higher than in 1957, coke consumption by blast furnaces was
15 percent *lower* in 1970 than in 1957.

---

[1]Using January 1970 prices, coke accounts for about 15 percent of
the price of steel. The prices used in these comparisons are those of
foundry coke and billet steel (rerolling, carbon in 1970; and forging,
carbon in 1976). The source was U.S. Department of Labor, Bureau of
Labor Statistics, *Wholesale Prices and Price Indexes for January 1970*
(May 1970), pp. 10, 15, and *Wholesale Prices and Price Indexes Data for
June 1976* (September 1976), pp. 22, 32. The computation assumed 0.633
tons of coke per ton of pig iron and 0.70 tons of pig iron per ton of
steel ingot, the 1970 averages.

In 1969, the steel industry consumed about 90 percent of total U.S. coke consumption, using all but a negligible fraction (0.4 percent) in blast furnaces. Coke consumed by other industries has been decreasing and seems unlikely to grow. Future demand for coal for coking will, therefore, depend almost entirely on use by the steel industry. There are three main links between steel production and the consumption of coal for coking purposes.

First, there is the amount of coal required per ton of coke. Although this ratio can, in principle, vary widely according to the physical properties of the coal used, it has not in fact done so, perhaps as a result of the blending of different types of coal. Coke output as a percentage of coal input has ranged between 70.3 and 69.4 during the postwar period, and it is unlikely to move much outside these bounds by 1980, as long as the steel companies use similar coal from their traditional sources.

Second, there is the amount of coke used per ton of pig iron. This ratio has declined from .95 in 1947 to .626 in 1969, although it has not declined in every year. The function of coke as support for the charge of iron ore, manganese, and other materials cannot be economically performed by any other material, but a series of small technological changes in blast furnace operations has gradually reduced coke requirements. Among the changes have been better sizing of the burden, lower coke ash and higher iron content of the burden, higher hot-blast temperatures, prefluxed burdens, and tuyere injected fuels.[1] A process of

---

[1]These factors were noted by James C. Gray, "Coal and the Steel Industry", presented at National Coal Association Technical-Sales Conference and Bituminous Coal Research, Inc., Annual Meeting, September 14-15, 1966, Pittsburgh, Pennsylvania.

pelletizing iron ore for electric furnace use, although not yet proven commercially, would, if successful, lead to a drastic reduction in the demand for coal for coking, since the process uses no coke at all. Since the first plants were started only in the late 1960's, it is too early to know the outcome of the venture.

In the projections for coke demand, three different decline rates in the coke-to-pig-iron ratio are used. All three are fairly close, however, so that not much of the difference between the high and low forecasts is due to this variation. All three rates were estimated by fitting a semi-logarithmic time trend to historical data.

Inspection of past values of this ratio showed that the average annual decline rate was essentially the same between 1949 and 1957 and between 1963 and 1969. The decline rate fell more sharply between 1958 and 1962. The three rates used were: (a) -1.2 percent, the average rate between 1963 and 1969; (b) -1.3 percent, the average rate for 1949-1957 and 1963-1969; and (c) -1.65 percent, the average rate for 1949-1969 allowing for downward shifts in the time trend in 1958 and 1963. These three decline rates, starting from a base of 62.6 percent in 1969, imply coke-to-pig-iron ratios in 1980 of .55, .54, and .52, respectively. Although these ratios are higher than those of another projection, they seem more in line with recent experience.[1] In any event, the effect of the differences on the projections is small, on the order of 10 percent of coke demand, and the effect relative to total demand for Appalachian coal is small indeed.

---

[1] A projection made by the European Communities, *Report on the Question of Coking Coal and Coke for the Iron and Steel Industry of the Community* (Brussels: 1969), p. 53, estimated the 1980 ratio to lie between .46 and .50.

The third ratio, pig iron consumption to steel production, has remained remarkably stable, ranging between 64 and 70 percent from 1947 to 1969, with an average value of 67 percent. This stability is remarkable in view of the radical changes in steelmaking technology over this period. Of the many innovations, the shift from the traditional basic open hearth to basic oxygen and electric furnaces has been only the most conspicuous. While the basic oxygen process uses relatively more pig iron than does the open hearth furnace, the electric furnace can, and often does, use almost exclusively scrap.

The apparent explanation for the stability in the pig-iron-to-steel ratio is that the price of scrap fluctuates so that virtually all scrap produced gets consumed. Most scrap used is "new" scrap, generated in the steel production process itself. In the past, the scrap market has adjusted in two ways. First, open hearth furnaces have varied the proportions of scrap and pig iron in the charge in response to changes in scrap prices. Second, a longer-run adjustment took place when, in the period of depressed scrap prices following the upsurge in basic oxygen furnace production, electric furnace production increased dramatically.

In the coming decade, the decline in importance of open hearth furnaces will lessen the impact of this form of adjustment. The relative output of electric furnaces and basic oxygen furnaces will still vary in response to sustained changes in the price of scrap. When scrap prices stay low, for example, existing electric furnace capacity is used more fully, and

125

eventually more electric furnace capacity is built. In addition, it is quite possible that, to increase flexibility, the steel industry will modify the basic oxygen furnaces to permit them to vary the proportions of the charge more readily.

In the short run, then, variations in the ratio of scrap to pig iron can and do occur before prices adjust to their equilibrium levels. In addition to the demand response, there may be some short-run supply response, with high scrap prices calling forth additional supplies of stockpiles of junked automobiles, for example, or low prices leading to additional stockpiling.

In the long run, however, the amount of scrap produced is essentially proportional to steel production, and the long-run supply of scrap is quite unresponsive to scrap prices. The ability of furnaces to vary the proportions of the charge, and the different proportions used by different types of furnaces ensure that, given a competitive market for scrap, virtually all of the scrap produced will be reused in steel production. On average, then, for the period to 1980 the ratio of pig iron consumption to steel production is likely to stay within its historically-observed range.

The forecasts for total coal consumed for coking purposes have two components -- one depending on coke used for steel, the other depending on the miscellaneous uses of coke. Because of its unimportance, coal use in the latter category was simply predicted to continue to decline at the annual average rate observed between 1948 and 1969, 3.3 percent.[1]

---

[1]This rate was estimated by a semi-logarithmic regression of other coke consumption as a function of time.

126

The predictions of coal used by the steel industry for coke for steel production depend, first, on predictions of steel output, and, second, on the ratios discussed above. The coke-to-pig-iron ratio was, alternately, assumed to decline at the three rates discussed. The coal-to-coke ratio was assumed to be .698, its postwar average, while the pig-iron-to-steel ratio was alternately assumed to be .64, .67, and .70, which represent the postwar low, the postwar average, and the postwar high, respectively. Finally, the ratio of coke produced to coke consumed was taken as 1.01, the postwar average.[1]

The resulting forecasts are shown in Table 5-1. The forecast gives a low, middle and high estimate. The low forecast, for example, uses the slowest rate of growth in GNP (and steel, which is taken to depend directly on GNP), the fastest decline in the coke-to-pig-iron ratio, and the lowest ratio of pig iron to steel output. The middle and high estimates are constructed analogously.

The low forecast shows a very gradually declining use of coal for coke by the steel industry, the middle forecast shows slow growth, while the high forecast (perhaps overly optimistic) predicts an annual average growth rate of 3.5 percent. The high and the low forecasts have been purposely constructed so as to bracket almost any reasonable development in the coking market. Even so, the difference between them is only about 7 percent of current coal production. The impact of this sector on the total market for coal, under the most extreme assumptions, is likely to be small.

---

[1]The divergence of this ratio from unity may reflect waste or simply a statistical discrepancy, but coke output exceeded coke input in 18 of the years from 1947 to 1969, so it seemed advisable to take it into account.

Table 5-1

FORECASTS OF CONSUMPTION OF COAL FOR COKING PURPOSES,
ASSUMING A DECLINE IN "OTHER" USES[1]

LOW

Year	GNP	Steel	Coal:Steel	Coal:Other	Coal:Coke
1970	724.300	126.973	72.4048	8.09761	80.5024
1971	749.125	128.226	71.9223	7.83475	79.7570
1972	774.801	129.536	71.4681	7.58042	79.0485
1973	801.356	130.917	71.0484	7.33435	78.3828
1974	828.822	132.385	70.6693	7.09626	77.7655
1975	857.229	133.954	70.3366	6.86591	77.2025
1976	886.610	135.640	70.0562	6.64303	76.6993
1977	916.998	137.459	69.8338	6.42739	76.2612
1978	948.428	139.428	69.6749	6.21875	75.8937
1979	980.934	141.564	69.5851	6.01688	75.6019
1980	1014.560	143.887	69.5695	5.82156	75.3910

MIDDLE

Year	GNP	Steel	Coal:Steel	Coal:Other	Coal:Coke
1970	724.300	126.973	76.0645	8.09761	84.1621
1971	751.601	129.025	76.2954	7.83475	84.1301
1972	779.931	131.205	76.5822	7.58042	84.1626
1973	809.329	133.531	76.9331	7.33435	84.2674
1974	839.835	136.022	77.3562	7.09626	84.4525
1975	871.491	138.699	77.8599	6.86591	84.7258
1976	904.340	141.583	78.4522	6.64303	85.0952
1977	938.427	144.696	79.1414	6.42739	85.5688
1978	973.800	148.060	79.9357	6.21875	86.1545
1979	1010.510	151.701	80.8435	6.01688	86.8603
1980	1048.590	155.643	81.8729	5.82156	87.6944

HIGH

Year	GNP	Steel	Coal:Steel	Coal:Other	Coal:Coke
1970	724.300	126.973	79.5499	8.09761	87.6475
1971	760.675	131.960	81.6879	7.83475	89.5227
1972	798.876	137.388	84.0338	7.58042	91.6142
1973	838.996	143.302	86.6054	7.33435	93.9398
1974	881.130	149.748	89.4214	7.09626	96.5177
1975	925.381	156.775	92.5009	6.86591	99.3668
1976	971.854	164.436	95.8636	6.64303	102.5070
1977	1020.660	172.785	99.5298	6.42739	105.9570
1978	1071.920	181.883	103.5210	6.21875	109.7400
1979	1125.750	191.792	107.8580	6.01688	113.8750
1980	1182.290	202.577	112.5650	5.82156	118.3860

[1]For units, explanation of column headings, and assumptions behind
the forecasts, see the continuation of the table.

Table 5-1 (Continued)
FORECASTS OF CONSUMPTION OF COAL FOR COKING PURPOSES,
ASSUMING A DECLINE IN "OTHER" USES

Units:

All amounts are in millions of short tons, except GNP,
which is measured in millions of 1958 dollars.

Column Headings:

GNP: Gross National Product of the United States.

STEEL: Production of steel ingots.

COAL:STEEL: Coal used for coking in the production of
                pig iron.

COAL:OTHER: Coal used for coking for other purposes than
                steel production.

COAL:COKE: Total coal used for coking, the sum of coal:
                steel and coal:other.

Assumptions:

Low Forecasts:

    a. GNP grows at an annual rate of 3.37 percent from
its 1970 level (724.3).

    b. The coke-to-pig iron ratio declines at an annual
rate of 1.65 percent from its 1969 level (.626).

    c. The ratio of pig iron production to steel ingot
production is .64.

Middle Forecasts:

    a. GNP grows at an annual rate of 3.7 percent from its
1970 level (724.3).

    b. The coke-to-pig iron ratio declines at an annual
rate of 1.3 percent from its 1969 level (.626).

    c. The ratio of pig iron production to steel ingot
production is .67.

Table 5-1 (Continued)

FORECASTS OF CONSUMPTION OF COAL FOR COKING PURPOSES,
ASSUMING A DECLINE IN "OTHER" USES

High Forecasts:

    a. GNP grows at an annual rate of 4.9 percent
from its 1970 level (724.3).

    b. The coke-to-pig iron ratio declines at an
annual rate of 1.2 percent from its 1969
level (.626).

    c. The ratio of pig iron production to steel
ingot production is .70.

All Forecasts:

    a. The ratio of coke output to coke input is 1.01.

    b. The ratio of coal input to coke output is 1.43.

    c. "COAL:OTHER" is assumed to decline at a rate of
3.3 percent per year from its 1969 level of
8.65008.

For a discussion of the equations used to forecast GNP,
steel production, and the time trends assumed, see the appendix
to this chapter.

130

Gordon's collection of other 1980 forecasts for coal for coking for blast furnaces (which corresponds to the columns labelled "Coal: Steel" in Table 5-1) shows forecast ranges which lie within the ones given here.[1] The other forecasts range from a low of 71.6 million short tons to 100.9 million short tons of coal, corresponding roughly to a range of 154 to 202 million short tons of steel. The forecasts shown in Table 5-1, therefore, bracket the existing forecasts, while the medium forecast is almost in the middle of the others.

In the time since these forecasts were made, six years of data have accumulated. In Table 5-2, the actual coal consumption used in coke production is compared with the low and high forecasts for 1970 through 1975. Actual coal consumption fluctuates much more widely from year to year than do the forecasts, which tend to grow or decline steadily. In one of the six years, the actual exceeded the high forecast (by less than 10 percent). Average actual coal consumption over this period was between the average for the high and the low forecasts, although closer to the high forecast. At least part of the divergence can be attributed to differences between the actual and forecasted levels of steel production, somewhat offset by a slower decline in the coal-to-steel ratio than forecasted. In all, however, the experience in the first half of the decade suggests that the forecasts in Table 5-1 will probably bracket the actual 1980 consumption levels.

---

[1]Gordon, (op. cit., Table 11-6) summarizes the forecasts of the European Communities (1969), Battelle (1969), and Occidental Petroleum (1971).

Table 5-2

COMPARISON OF FORECASTED WITH ACTUAL COAL
CONSUMED FOR COKE PRODUCTION, 1970-1975

(Millions of Tons)

Year	Low Forecast	High Forecast	Actual
1970	80.5	87.6	96.0
1971	79.8	89.5	82.8
1972	79.0	91.6	87.3
1973	78.4	93.9	93.6
1974	77.8	96.5	89.7
1975	77.2	99.4	83.3
Average	78.8	93.1	88.8

SOURCES:

Low and high forecasts -- Table 5-1.

Actual -- 1970-1973: U.S. Bureau of Mines, *Mineral Surveys*, "Coal--Bituminous and Lignite for 1973".

Actual -- 1974-1975: *Keystone News Bulletin*, Vol. 34, No. 4, compiled from Bureau of Mines, *Weekly Coal Report*.

Forecasts of demand for coking coal from Appalachia are shown in Table 5-3. Because about 90 percent of coking coal comes from Appalachia, the regional forecasts are very close to the total forecasts.

The most important variable in these forecasts, steel production, is difficult to predict, since it depends not only on general economic activity but also on foreign trade policy. The range shown here seems likely to bracket any probable events to 1980, but drastic import restrictions might, for example, push steel production toward the high end of the forecasts, while a laissez-faire policy would, given current market conditions, restrict steel's growth toward the low end.

An increase in steel imports might not, however, lead to a concomitant decrease in coking coal *production*, since exports of coal to Europe and Japan are largely for coking. Exports, considered below, depend primarily on domestic policies followed by the European countries and on coal prices in other countries such as Canada and Australia.

Exports of Coal

In 1975 the United States exported 65.5 million tons of coal, about 10 percent of domestic production. All but a negligible fraction (1.2 percent) of the exports were produced in the Appalachian region, with Districts 7 and 8 accounting for almost 70 percent of total exports (80 percent of overseas exports).[1] Domination of the export market

---

[1] These shares were computed from U.S. Bureau of Mines, *Bituminous Coal and Lignite Distribution,* Quarterly, April 12, 1976, the source of all origin and destination data reported in this section.

Table 5-3

FORECASTS OF COAL FOR COKING, ASSUMING A DECLINE IN "OTHER" USES

(Millions of Tons)

LOW

Year	Appalachian Coal	Total Coal
1970	72.07	80.50
1971	71.41	79.76
1972	70.77	79.05
1973	70.18	78.38
1974	69.62	77.77
1975	69.12	77.20
1976	68.67	76.70
1977	68.28	76.26
1978	67.95	65.89
1979	67.69	75.60
1980	67.50	75.39

MIDDLE

1970	75.35	84.16
1971	75.32	84.13
1972	75.35	84.16
1973	75.44	84.27
1974	75.61	84.45
1975	75.86	84.73
1976	76.19	85.10
1977	76.61	85.57
1978	77.13	86.16
1979	77.77	86.86
1980	78.51	87.69

Table continued on following page.

Table 5-3 (Continued)

FORECASTS OF COAL FOR COKING,
ASSUMING A DECLINE IN "OTHER" USES

(Millions of Tons)

HIGH

Year	Appalachian Coal	Total Coal
1970	78.47	87.65
1971	80.15	89.52
1972	82.05	91.64
1973	84.10	93.94
1974	86.41	96.52
1975	88.96	99.37
1976	91.78	102.51
1977	94.86	105.96
1978	98.25	109.74
1979	101.95	113.88
1980	105.99	118.39

by Appalachian coal results partly from the quality of the coal and partly from the region's location.

Much Appalachian coal is low in sulfur and ash, high in heat content, and will coke. Coal having these properties sells at a considerably higher price than ordinary steam coal.[1] Consequently, transportation costs are a relatively smaller part of the delivered price in the importing countries. On these grounds alone, then, the high-grade Appalachian coals are better able to compete with European coals than are, say, midwestern coals. In addition, West Virginia is close to the Eastern Seaboard, so that the costs of delivering coal to Hampton Roads for export are well below the transport costs from other regions.

In 1975 overseas exports accounted for about 48 million tons, with exports to Canada making up the rest (except for less than 0.6 million tons sent to Mexico). Almost all of the exports to Canada were Appalachian coal, with all but a negligible fraction being shipped on the Great Lakes. Shipments to Canada were primarily for coke plants and for electric utilities, in about equal proportions. The steam coal was shipped from Appalachian districts relatively close to the Great Lakes (Districts 2, 3 and 6, primarily), while the districts that shipped virtually all of the coal for coking (Districts 7 and 8) shipped negligible amounts of steam coal. This dichotomy illustrates the twin factors of coal quality and

---

[1]For example, in February 1976, utilities in West Virginia paid an average of $23.50 per ton for coal with less than 1 percent sulfur, while utilities in Illinois paid an average of $14.00 per ton for coal with more than 3 percent sulfur. While these figures may not reflect exactly the difference between coking coal and steam coal prices, they are reasonably representative.

locational advantage even within Appalachia, both of which are enjoyed by Districts 7 and 8 with respect to the overseas export market.

Exports during the postwar period fluctuated widely. The fluctuations have largely resulted from international political occurrences or from the domestic coal policies of European Economic Community (EEC) countries.[1] Total exports for 1950 to 1970, with subtotals for EEC countries, Canada and Japan shown separately, are shown in Table 5-4. The large increase in exports in 1956-1957 was the result of the Suez crisis, and the high rates of U.S. exports to EEC countries in those years led, in 1958, to policies in those countries which favored domestic coal at the expense of imports from the United States. After the initial fall-off, however, exports from the United States to EEC countries began growing again in the early 1960's, until a second decline set in in the mid-1960's, mainly as a result of European Coal and Steel Community subsidies to the coking coal industry.[2] Even these subsidies appear inadequate to encourage sufficient domestic production in those countries, given the sizable increase in imports from the United States in 1970.[3]

---

[1]An authoritative account of the postwar European coal market appears in Richard L. Gordon, *The Evolution of Energy Policy in Western Europe: The Reluctant Retreat from Coal* (New York: 1970). The historical discussion in this section relies on this source and on Gordon's manuscript on the demand for coal, *op. cit.*, which is the basis for the analysis of probable future events in this market.

[2]Gordon, manuscript on coal demand, *op. cit.*, p. XI-8.

[3]*Ibid.*

Table 5-4

U.S. COAL EXPORTS, 1950-1974

(Millions of Tons)

Year	Canada	European Economic Community Countries[1]	Japan	Total[2]
1950	23.0	0.2	0.1	25.5
1951	23.8	20.0	1.6	56.7
1952	21.0	16.3	2.9	47.6
1953	19.6	7.3	3.9	33.8
1954	15.9	7.2	2.9	31.0
1955	17.2	19.5	2.8	51.3
1956	20.7	32.8	3.2	68.6
1957	18.4	41.7	4.9	76.4
1958	12.2	27.5	3.3	50.3
1959	12.4	15.1	4.0	37.3
1960	11.6	14.0	5.6	36.5
1961	11.2	12.9	6.6	35.0
1962	11.4	15.6	6.5	38.4
1963	13.8	21.4	6.1	47.1
1964	14.2	21.1	6.5	48.0
1965	15.7	21.3	7.5	50.2
1966	15.8	19.3	7.8	49.3
1967	15.3	16.3	12.2	49.5
1968	16.7	12.0	15.8	50.6
1969	16.8	11.9	21.4	56.2
1970	18.7	16.7	27.6	70.9
1971	17.6	12.8	19.7	56.6
1972	18.2	13.5	18.0	56.0
1973	16.2	10.7	19.2	52.9
1974	13.7	13.0	27.3	59.9

Footnotes and sources on following page of table.

Table 5-4 (Continued)

U.S. COAL EXPORTS, 1950-1974

[1]For 1950-1959, includes Luxembourg, France, West Germany, Italy, and Netherlands.

[2]Includes exports to countries other than those shown separately.

SOURCES:

1950-1970:   U.S. Bureau of Mines, *Minerals Yearbook,* various years.

1971-1974:   U.S. Bureau of Mines, *Mineral Industry Surveys,* "Coal--Bituminous and Lignite in 1973" and "Coal-- Bituminous and Lignite in 1974" (January 1975 and January 1976).

The transportation costs for coal, coupled with the low prices
of residual fuel oil in Europe, have led to a rapid dwindling of coal
exports for nonmetallurgical purposes.[1] In 1970, for example, 75 percent
of coal exported to Europe was for metallurgical purposes (coking), up
substantially from about 50 percent in the early 1960's.[2] The future
for U.S. coal exports to Europe, therefore, depends almost entirely on
the market for metallurgical coal.

Exports to Japan, which have increased rapidly along with the fast
growth in that country's steel output, are entirely for metallurgical
use. Japan does not restrict coal imports, however, so that the level
of U.S. exports depends (for a given grade of coal) solely on the
delivered prices of U.S. coals relative to those of its competitors,
primarily western Canada and Australia. Transportation costs from
the latter countries, however, are substantially lower than those from
East Coast ports.[3]

About 53 percent of U.S. coal exports to Canada are destined for
electric utility consumption, 42 percent for coking, and the remaining
5 percent for miscellaneous purposes.[4] This pattern is markedly different

---

[1] *Ibid.*

[2] U.S. Bureau of Mines, *International Coal Trade,* monthly.

[3] Gordon, manuscript on coal demand, *op. cit.,* p. XI-13.

[4] U.S. Bureau of Mines, *Bituminous Coal and Lignite Distribution,*
April 12, 1976, p. 42.

from that for overseas exports. Indeed, because of geographical proximity and traditional trade patterns, exports to Canada are more naturally analyzed, perhaps, within the framework for U.S. domestic shipments. Transportation costs from the United States to Canada, the delivered price of western Canadian coal in the eastern provinces, and the level of general economic activity in Canada are the major determinants of U.S. exports to this country.

Future developments in these three principal export markets -- the European Economic Community countries, Japan and Canada -- depend on factors specific to each, as well as on the general level of U.S. coal prices.

Forecasts of U.S. Coal Exports

Gordon discusses each country's probable demand for U.S. coal.[1] Because a full treatment can be found there, I summarize here, in Table 5-5, the range of forecasts of U.S. exports. The range, between 68 and 129 million tons, is wide, but it does bracket other forecasts.[2] The generally stable level of exports from 1970 to 1975 suggests that the 1980 level may lie between the low and the middle forecasts. In any case, the difference is unlikely to account for a significant share of Appalachian output.

---

[1]Gordon, Chapter XI.

[2]Occidental Petroleum forecast 107 million tons, while the National Petroleum Council forecast 110 million tons.

141

Table 5-5

SUMMARY OF 1980 FORECASTS OF U.S. COAL EXPORTS

(Millions of Short Tons)

Market	Low	Middle	High
EEC Countries	20	27	35
Japan	19	36	51
Canada	19	22	24
Others	10	15	19
Total	68	100	129

Retail Demand

Sales by retailers to consumers are primarily for home heating. At
one time among the largest end uses of coal, retail dealer deliveries
plummeted in the years after World War II, as consumers converted their
furnaces to cleaner and more convenient fuels such as natural gas and
distillate fuel oil.

Although the decision to convert a furnace is influenced by relative
fuel prices, coal consumption in this category has fallen so steadily
to such a low level (1 percent of total consumption in 1975) that any
estimated price effects would have only a negligible impact on total coal
consumption.

Therefore, a simple exponential decline rate was used to forecast
consumption in this end use. The semi-logarithmic time trend, estimated
by ordinary least squares, implied an average yearly decline of 9.4
percent between 1949 and 1969:

$$\log RC = 4.56 - .094\ T \tag{12}$$
$$(153.6)\ (-35.1)$$

$R^2 = .985$ $\qquad\qquad R(C)^2 = .983$

Standard error of regression $= .074$

Observations: 1949-1969

where RC is retail dealer deliveries, in million of tons, and T is time,
with 1949 = 1.

If the historical decline rate continues, consumption in this category will fall from 12.7 million tons in 1969 to 4.52 million tons in 1980. Estimated consumption of Appalachian coal in this category will decline from 9.41 million tons in 1969 to 3.35 million tons in 1980.

In Table 5-6, actual and forecasted retail consumption are compared for 1970 to 1975. Although actual consumption fluctuates more than the forecast, the predicted decline seems approximately correct.

"Other Manufacturing and Mining" Demand for Coal

The category "other manufacturing and mining" uses of coal, as reported in Bureau of Mines publications, is a heterogeneous grouping of industries using coal primarily for the generation of process steam.[1] This section discusses coal use in two other categories as well. Although "steel and rolling mill" use and "cement mills" use are reported separately by the Bureau of Mines, we will consider them together.

"Other manufacturing and mining" is a relatively large category of consumption, accounting in 1975 for about 11 percent of the total. Coal use in this category has varied somewhat in response to changes in the general level of economic activity, but there has been no apparent response to changes in relative fuel prices. This insensitivity is

---

[1]Electricity is usually a joint product, and some industrial consumers generate substantial amounts.

Table 5-6

ACTUAL AND FORECASTED RETAIL COAL CONSUMPTION,
1970-1975

(Millions of Tons)

Year	Actual	Forecast
1970	12.1	11.5
1971	11.4	10.5
1972	8.7	9.6
1973	8.2	8.7
1974	8.8	7.9
1975	5.7	7.2

SOURCE: "Actual" -- U.S. Bureau of Mines, *Mineral Industry Surveys*, "Coal--Bituminous and Lignite in 1973", and "Weekly Coal Report", April 9, 1976, p. 8.

somewhat surprising. Although industries using coal to make process steam would find it disproportionately expensive to maintain capacity to burn more than one fuel, conversion to oil is relatively inexpensive. The explanation may be that the coal-burning manufacturing plants are located away from the eastern seaboard, so that they did not have access to imported residual fuel oil when its price was low relative to that of coal.

In any case, no attempt was made to sort out the myriad minor influences on this diverse grouping. Instead, coal use in this category was regressed by ordinary least squares on a time trend and on the Federal Reserve Board index of durable manufactures production. Because consumption in this category varied almost randomly between 73 and 93 million tons over the period 1955 to 1969, it is hardly surprising that the estimated equation fits rather poorly:

$$\log OC = \underset{(2.83)}{1.881} + \underset{(3.851)}{0.645} \log YDUR - \underset{(-3.99)}{.033} T \tag{13}$$

$R^2 = .57 \qquad\qquad R(C)^2 = .46$

Standard error of equation $= .047$

Observations: 1955-1969

where OC is coal use in the "other manufacturing and mining" category, $YDUR$ is the index of durable manufacturing, and T is a time trend with 1955 = 1.

146

The three forecasts of consumption in this category, which differ in the assumption made about the growth rate of Gross National Product, are shown in Table 5-7. The difference between the high and the low forecast values for Appalachian coal use in this category in 1980 is only about 7 million tons, a negligible amount relative to total Appalachian production.

Coal used in "steel and rolling mills" is separate from the steel industry's use of coal for coking, and it is a small and declining end use for coal. Around 11 million tons in 1950, consumption in this category was less than 3 million tons in 1975, or about 0.5 percent of total consumption in that year. Coal consumption by cement mills, on the other hand, has stayed fairly stable, ranging between 7.6 and 9.4 million tons between 1955 and 1969. The sum of these two categories has varied only between 14.7 and 16.3 million tons over the same period. Because any other reasonable procedure would yield negligibly different results, no appreciable error will be made in assuming that the sum of consumption in these two categories will remain at 15 million tons over the forecast period.

Summary

In this chapter I discussed and forecasted consumption of coal in non-utility uses. As demand in these uses is relatively price inelastic, I have used a single set of these numbers for the different scenarios presented in Chapters 6 and 7. The 1980 forecasts of

147

Table 5-7

FORECASTS OF COAL CONSUMED IN OTHER MANUFACTURING AND MINING

(Millions of Tons)

LOW

Year	Appalachian Coal	Total Coal
1970	58.01	82.29
1971	57.62	81.75
1972	57.24	81.20
1973	56.86	80.66
1974	56.48	80.12
1975	56.10	79.59
1976	55.73	79.06
1977	55.36	78.54
1978	54.99	78.01
1979	54.63	77.49
1980	54.26	76.98

MIDDLE

Year	Appalachian Coal	Total Coal
1970	58.01	82.29
1971	57.77	81.96
1972	57.53	81.62
1973	57.30	81.29
1974	57.06	80.95
1975	56.83	80.62
1976	56.60	80.29
1977	56.37	79.97
1978	56.14	79.64
1979	55.91	79.31
1980	55.68	78.99

Table continued on following page.

148

Table 5-7 (Continued)

FORECASTS OF COAL CONSUMED IN OTHER MANUFACTURING AND MINING

(Millions of Tons)

HIGH

Year	Appalachian Coal	Total Coal
1970	58.01	82.29
1971	58.32	82.73
1972	58.62	83.17
1973	58.93	83.61
1974	59.25	84.05
1975	59.56	84.49
1976	59.87	84.94
1977	60.19	85.39
1978	60.51	85.84
1979	60.83	86.29
1980	61.15	86.75

non-utility coal consumption are shown in Table 5-8. They generally correspond to the middle forecast where a range has been shown, and they have been distributed among Appalachian producing districts according to the 1970 shipments pattern for each end use.

As a benchmark, the electric utility consumption shown is the simple extrapolation, assuming no competition with residual fuel oil. Total shipments from Appalachia in 1980 are shown to be around 521 million tons, of which over 280 million tons are consumed by electric utilities. In Chapters 6 and 7, electric utility consumption is analyzed more carefully in the light of sulfur emission restrictions and the 1973-1974 OPEC oil price increases.

Table 5-8

SUMMARY OF 1980 BASE FORECASTS OF COAL CONSUMPTION
BY APPALACHIAN PRODUCING DISTRICT

(Millions of Tons)

District	Coking	Exports[1]	OTHER			Electric Utility[2]	District Total
			Retail	Other	Total Other		
1	4.67	5.41	0.21	6.25	16.53	48.25	64.78
2	22.89	3.72	0.08	5.17	31.87	11.64	43.50
3 & 6	5.07	10.93	0.08	4.73	20.81	55.16	75.97
4	0.00	0.72	0.46	11.14	12.33	67.71	80.04
7	14.20	26.41	0.20	1.60	42.40	1.48	43.89
8	27.27	52.11	2.16	24.36	105.90	82.83	188.73
13	6.22	0.00	0.12	2.04	8.39	16.00	24.38
TOTAL	80.32	99.30	3.31	55.29	238.23	283.07	521.30

NOTE: Detail may not add to totals because of independent rounding.

[1] Includes shipments to Canada.

[2] Assumes that no markets are lost to residual fuel oil.

Appendix 5A

FORECASTING EQUATIONS AND TIME TRENDS

No attempt was made to estimate the underlying structure, and ordinary least squares was used in every case. This procedure seemed sensible in view of the historical regularity of many of the relationships.

Economy-Related Variables

The driving variable in coal consumption for coking by the steel industry is the production of steel ingots. Steel production, in turn, is closely related to the manufacture of durable goods. The estimated equation was:

$$S = 14.41 - .099 \; DSTR - .137 \; T^2 + 1.077 \; YDUR \qquad \text{(A.1)}$$
$$(0.365) \; (-2.01) \qquad (-4.16) \quad (6.81)$$

YEARS = 1947-1968   Durbin-Watson = 1.54

$R^2 = .883$ $R^2(C) = .857$

The numbers in parentheses below the coefficients are *t-statistics*, R^2 is the coefficient of multiple determination, $R^2(C)$ is that

152

coefficient adjusted for degrees of freedom, and the Durbin-Watson statistic is a measure of autocorrelation of the residuals. This nomenclature will be used throughout this section.

The variables in equation (A.1) are:

S: total production of steel ingots, in millions of short tons;

DSTR: the number of days lost in general steel strikes (assumed to be zero throughout the forecast period);

T: a time trend, with 1947 = 1;

YDUR: Federal Reserve Board index of durable manufactures, 1957-1959 = 100.

Equation (A.1) shows steel production increasing with durable manufacturing, but with a secular decrease, perhaps related to growing competitiveness of Japanese and European steel.[1]

Forecasting equation (A.1) requires a forecast of *YDUR*, which is closely related to general economic activity as measured by real gross national product. The equation for *YDUR* is:

$$\log YDUR = -2.793 + 1.211 \log GNP \qquad\qquad (A.2)$$
$$(-8.30)\ (30.98)$$

YEARS = 1947-1969   Durbin-Watson = 1.44

$R^2 = .979$ $R^2(C) = .977$

The fit is very good, so that we would expect predictions of *YDUR* to be reasonably accurate.

---

[1]A time trend term in this equation was included in this equation solely for forecasting purposes. Equation A.1 is not intended as a structural equation explaining steel production.

It remains only to forecast *GNP* (gross national product evaluated at 1958 prices). The growth of *GNP* is very closely related to government policies, however, and it has varied considerably over the postwar period. Because of this variability it was decided to estimate three different growth rates based on postwar experience. The period from 1947 to 1961 was a period of relatively slow growth, and the estimated equation gives the smallest coefficient of the time term:

$$\log GNP = 5.737 + .0337\ T \tag{A.3}$$
$$(15.89)\quad(15.69)$$

YEARS = 1947-1961   Durbin-Watson = .842

$R^2 = .950$ $R^2(C) = .942$

where 1947 is year 0, and the logarithm is in the natural base.

From 1961 to 1969 the economy, first recovering from a recession and then spurred by the Vietnam War, showed an exceptionally high rate of growth, probably in excess of its long-run capabilities. For this reason it serves as an upper bound on potential economic growth in the 1970's. The estimated equation was:

$$\log GNP = 5.490 + .0487\ T \tag{A.4}$$
$$(25.10)\ (27.84)$$

YEARS = 1961-1969   Durbin-Watson = 1.42

$R^2 = .991$ $R^2(C) = .989$

The average growth rate of the economy from 1947 to 1969 was taken as the most probable average growth rate for the economy between 1970 and 1980. The estimated equation was:

154

$$\log GNP = 5.713 + .037\ T \tag{A.5}$$
$$(32.0)\quad (33.5)$$

YEARS = 1947-1969 $\qquad\qquad$ Durbin-Watson = .653

$R^2 = .982$ $\qquad\qquad\qquad\qquad$ $R^2(C) = .980$

Time Trends

The uneven decline rate of the coke-to-pig iron ratio over the postwar period, with its slow decline from 1949 to 1957 and from 1963 to 1969, and its rapid decline from 1958 to 1962, led to several estimated equations. The first, reflecting only the most recent behavior of the rate, was:

$$\log R = 4.372 - .012\ T \tag{A.6}$$
$$(56.9)\quad (-2.81)$$

YEARS = 1963-1969

$R^2 = .612$ $\qquad\qquad\qquad\qquad$ $R^2(C) = .457$

where T in 1949 equals 1, and where R is the coke-to-pig iron ratio, in percentages.

The average decline rate for the years 1949 to 1957 and 1963 to 1969 was estimated by the following equation:

$$\log R = 4.395 + .164\ C1 - .013\ T \tag{A.7}$$
$$(137.2)\ (6.72)\quad\ (-7.53)$$

YEARS = 1949-1957, 1963-1969

$R^2 = .992$ $\qquad\qquad\qquad\qquad$ $R^2(C) = .991$

and where $C1$ is a dummy variable allowing the time trend to shift down between 1957 and 1963. That is, $C1$ has the value 1 for the years 1949 to 1957 and 0 elsewhere.

155

A similar equation was estimated for the entire period, allowing shifts between 1957 and 1958 and between 1962 and 1963. This equation imposes an identical decline rate for the whole period, although it allows the level of the years 1958 to 1962 to differ from both the period before and the period after it. The estimated equation was:

$$\log R = 4.454 - .0165\ T + .121\ C1 + .049\ C2 \qquad \text{(A.8)}$$
$$\quad\ \ (85.7)\ \ \ (-5.81)\ \ (3.05)\ \ \ \ (2.06)$$

YEARS = 1949-1969

$$R^2 = .973 \qquad\qquad\qquad R^2(C) = .966$$

where the other variables are as before and $C2$ is a dummy for the years 1958 to 1962.

The equation used to estimate the trend in nonsteel industry uses of coke was:

$$\log NSIC = 9.068 - .033\ T \qquad\qquad \text{(A.9)}$$
$$\qquad\quad\ \ (115.4)\ (-4.25)$$

YEARS = 1948-1969

$$R^2 = .488 \qquad\qquad\qquad R^2(C) = .434$$

where $NSIC$ is nonsteel industry uses of coke, in millions of short tons, and T equals 1 in 1948. To obtain the coal required to produce this amount of coke, the coke consumption was multiplied by 1.44, the very stable ratio prevailing in the postwar period.

156

Chapter 6

SULFUR EMISSION RESTRICTIONS

In this chapter, I use the model developed in Chapter 5 to determine the impact of different sulfur emission standards on coal output and employment in Appalachia. This analysis is based on the assumption that relative coal and oil prices remain at their 1969 level. The forecasts, therefore, describe what would have happened to Appalachian coal production if OPEC had not successfully raised oil prices.

The chapter is divided into two sections. The first sets forth the methods used to estimate Appalachian mining employment and sulfur emission control costs. The second presents forecasts under a 0.7 and a 2.0 percent sulfur-in-fuel standard.

Methods of Estimating Employment and Costs

Employment

The method used here to translate output by Appalachian producing district into employment by district is discussed in detail in the

Charles River Associates report.[1] The assumptions and analysis are
only sketched here. Although they seem reasonable, they necessarily
include arbitrary elements. Other assumptions about productivity increases
or relative growth of strip mining would affect the estimate of
output per man-day in the Appalachian producing districts. If, as
seems reasonable, the same output per man-day were applied to the
output from each of the scenarios, then the percentage change in
employment from the base scenario would not be affected. Consequently,
the estimate of absolute levels of employment is much more sensitive
to the assumptions used here than is the relative impact on employment
of the different policies.

Coal production techniques are usually classed as either surface
(or strip) or underground mining. Strip mining is much more capital
intensive, with much higher output per man-day, than underground.
Thus the job of forecasting output per man-day by district falls into
two parts: first, forecasting the output per man-day for each
technique; second, forecasting the share of output that will be
produced by each technique.

Output per man-day by technique was assumed to grow, after 1971,
at the same rate as that before 1969. Because of the provisions of
the Federal Coal Mine Health and Safety Act, it is assumed that output

---

[1]Charles River Associates, Incorporated, Volume I, pp. 378-380
and Appendix 8-A (pp. 388-406).

per man-day remains constant over the 1969-1971 period.[1] These

assumptions lead to a forecast of output per man-day in 1980 of

27.0 tons in underground mining and 54.0 tons in strip mining.[2]

Except in Ohio, strip output has, over time, been increasing relative

to underground output in all of the Appalachian coal producing districts.

Charles River Associates found that the ratio of strip to underground

output could be explained by its value 10 years before, by an index

of wages in bituminous coal mining, and by the wholesale price index

for general purpose machinery and equipment. Using this equation

and then forecasting the wage and rental variables by time trends,

they estimated that the 1980 value of this ratio would be 1.76 times

its 1970 value.[3] The 1980 value for District 4 (Ohio) was assumed

to be the same as the 1970 value, since the strip to underground ratio

has shown no historical trend.[4] As each district in 1970 had a

---

[1]*Ibid.*, p. 380.

[2]These assumptions have proven wildly optimistic. Output per
man-day in underground mining, on average, declined from 15.6 tons in
1969 to 8.5 tons in 1976, while productivity in strip mining stayed
roughly constant at 36 tons per man-day between 1969 and 1974 but fell
to 26 tons per man-day in 1975 and 1976. (U.S. Bureau of Mines, *Minerals
Yearbook* and preliminary releases of "Coal--Bituminous and Lignite in
1975".) The implication of the fall in productivity is that the
forecasts of employment will be substantially understated. The percentage changes in employment (as deviations from the base forecast)
resulting from the different policies will not, however, be affected.

[3]*Ibid.*, p. 404.

[4]*Ibid.*, p. 406.

different ratio of strip to underground production, the 1980 ratios were also predicted to be different. Output per man-day by technique was assumed to be the same for all districts, so that only the split between strip and underground accounted for the different forecasts of output per man-day in the different producing districts.

The calculation of the forecast of 1980 employment (in man-days) by district can be summarized in symbols as follows:

$$E_i = \frac{Q_{ui}}{27.0} + \frac{Q_{si}}{54.0} \qquad (14)$$

where:

E_i is 1980 employment in Appalachian district i;

Q_{ui} is 1980 underground production in district i;

Q_{si} is 1980 strip production in district i.

Production by technique is derived as follows:

$$Q_{ui} = \frac{Q_i}{1 + 1.76\ R_i} \qquad (15)$$

$$Q_{si} = (\frac{1.76\ R_i}{1 + 1.76\ R_i})\ Q_i \qquad (16)$$

where:

Q_i is 1980 output in district i;

R_i is the 1970 ratio of strip to underground output, district i.

For ease of computation, the formula for E_i can be expressed as:

$$E_i = a_i Q_i \qquad (17)$$

160

where:

$$a_i = \frac{1}{(1 + 1.76R)\ 27.0} + \frac{1.76R}{(1 + 1.76R)\ 54.0} \tag{18}$$

The value of a_i above translates tons of coal into man-days of employment. The results of this chapter are discussed in units of millions of tons of coal and man-years, where it is assumed that there are 240 man-days in a man-year. If a_i denotes the factor that translates millions of tons of output for the i-th district into man-years of employment for that district, the values of a_i for the different districts are shown in Table 6-1.

Even given employment in the Appalachian coal industry, it is difficult to assess the relative importance of this employment. Figures on total employment in Appalachia (as distinct from employment in states that include parts of Appalachia) are difficult to obtain. Table 6-2 compares coal mining employment with total employment in 1970 for the Appalachian states producing significant amounts of coal. Although the total coal mining employment is less than 1 percent of total employment, this share does not reflect the concentration of coal production in some counties or the isolation of these communities from the rest of the state. Even the share of employment in West Virginia directly accounted for by coal mining (about 9 percent) understates the importance of coal in particular communities.

Costs of Sulfur Emissions Control

The discussion of the methods used to estimate the costs of sulfur emissions control is organized by the type of fuel planned for

161

Table 6-1

MAN-YEARS PER MILLION TONS OF COAL,
BY APPALACHIAN COAL PRODUCING DISTRICT, 1980

District	Strip Output, 1970 (Millions of Tons)	Underground Output, 1970 (Million Tons)	Man-Years Per Million Tons, 1980 (a_i')
1	21.36	25.40	108.273
2	4.78	34.19	139.084
3 & 6	5.86	43.77	139.607
4	32.62	17.98	104.580
7	4.59	32.64	139.013
8	34.27	126.42	129.398
13	8.84	11.23	109.507

SOURCES:

Strip and Underground output -- Charles River Associates, Incorporated, *op. cit.*, Volume I, Table 8-1, pp. 351-357.

The factors were calculated according to the following formula (discussed in the text):

$$a_i' = (10^6/240) \cdot \left(\frac{1}{(1 + 1.76R_i) \cdot 27.0} + \frac{1.76R_i}{(1 + 1.76R_i) \cdot 54.0}\right)$$

where R_i = Strip output/Underground output, district i, 1970.

Table 6-2

EMPLOYMENT IN APPALACHIAN STATES, 1970

(Thousands)

State	Total Non-Agricultural	Bituminous Coal and Lignite Mining - SIC 12
Ohio	3880.7	9.3
West Virginia	516.7	45.3
Alabama	1010.4	5.3
Pennsylvania	4347.3	24.1
Tennessee	1327.6	2.0
Virginia	1465.3	12.1
Kentucky	910.3	20.5[1]
TOTAL	13458.3	118.6

[1]Eastern portion only

SOURCE: U.S. Department of Labor, Bureau of Labor Statistics, *Employment and Earnings, States and Areas 1929-71* (Bulletin 1370-10, 1974). Tennessee coal mining employment figure taken from National Coal Association, *Bituminous Coal Data,* 1971 edition, p. 30. *Ibid.,* for Kentucky Eastern.

and chosen by the plant. The parameters and method vary somewhat across the different policies examined, but the underlying principle is the same in each case: the costs of sulfur emissions control are the difference between the costs of burning fuel in compliance with the sulfur restriction and the fuel costs that would have been incurred without that policy. (The costs of any pollution control devices are implicitly included in fuel costs in this statement.) Since utilities are assumed to minimize the as-burned cost of fuel, subject to the sulfur standards, the cost of control reflects the difference between the cost of the unconstrained choice and the cost of the cheapest alternative to meet the constraint.

Plants Planned To Burn Coal That Choose Coal

The control costs at these plants are the easiest to compute, being the cost per ton times the number of tons burned. For the 0.7 percent sulfur-in-fuel standard, the cost per ton is assumed to be $5.00, which includes both the capital and the operating costs of stack gas desulfurization.[1]

---

[1]This estimate of the costs of stack gas desulfurization is taken from Charles River Associates, *op. cit.*, Volume II, pp. 86-99. At the time this estimate was made, no process had been commercially proven, and this estimate represents a reasonable guess at utilities' perceptions of these costs at that time. As discussed in Chapter 7, more recent estimates of these costs cover a wide range, both because of divergent estimates of capital and operating costs and because of differences among plants (even assuming the same capital cost; that is, the same capital cost faced by a plant with 30 years' remaining life implies quite a different cost per ton of coal than that faced by a plant with 5 years' useful life). For the age structure of plants potentially able to convert from oil to coal, the stack gas desulfurization costs ranged from $5.30 to $11.90 per ton (assuming the lowest estimates of stack gas operating and capital costs) or from $20 to $35 per ton (assuming the highest estimates of stack gas operating and capital costs).

For the 2.0 percent sulfur-in-fuel standard, it is assumed that Appalachian coal can be washed to meet the standard for $0.75 per ton, while stack gas desulfurization will be required for the higher sulfur midwestern coals, at a cost of $4.00 per ton.[1] For plants in Illinois, Indiana and Wisconsin that switch from midwestern coal to Appalachian coal, the cost per ton is the difference in the delivered price of Appalachian coal (including washing) and the 1969 as-burned cost of coal at the plant, adjusted to include FCMHSA costs.

Plants Planned To Burn Oil

Since the control costs associated with coal are higher than those associated with oil at every sulfur-in-fuel standard, a plant that was planned to burn oil even before the sulfur constraint had to be met would, of course, continue to burn oil. Different sulfur-in-fuel standards impose different control costs.

For the 0.7 percent sulfur-in-fuel standard, the cost is simply the number of barrels of oil times $0.75 per barrel, which is assumed to be the cost of desulfurization. In the analysis of the 2.0 percent sulfur-in-fuel standard, it is assumed that there is no additional cost of obtaining residual fuel oil with that sulfur content.[2]

Plants Switching From Coal To Oil

For plants that were predicted by the model to switch from coal to fuel oil anyway, the control costs were assumed to be the same as if the plant were planned to burn oil (as discussed in the preceding

---

[1]*Ibid.*, pp. 98-99.

[2]See Chapter 4 for a discussion of these costs.

section). For plants that would, in the absence of sulfur standards, have burned coal, the cost was the difference between the delivered as-burned price of residual fuel oil computed by the model (including the costs of desulfurization to the sulfur-in-fuel standard in question) minus the 1969 as-burned cost of coal (taken to include the $0.75 per ton due to the FCMHSA). These costs, measured in cents per million BTU's, were applied to the BTU content of the coal that would have been burned to arrive at the total cost at each plant.

Output, Employment and Costs

In this section, I present the forecasts of coal output and employment in Appalachia and sulfur emission control costs for a number of different assumptions. All of these forecasts are based on the assumption that relative coal and oil prices remain at their 1969 level. The forecast derived by simple extrapolation implicitly assumes that coal maintains its share of fossil fuels burned by electric utilities and that there is no sulfur-in-fuel standard in effect. The three other forecasts are based on the assumption that utilities choose the least-cost fuel and, implicitly, that the increased fuel costs do not lead to a decrease in electricity consumption. The sulfur-in-fuel standards analyzed are: a 0.7 percent standard, a 2.0 percent standard, and no standard at all. In all cases, the calculations have been made for the assumptions of low and high residual oil transport costs.

166

As shown in Chapter 4, the difference between the two sets of assumptions about transport costs is not large at 1969 relative prices and without a sulfur-in-fuel standard (the high transport costs and prices predict Appalachian coal output only about 7 percent higher than the output predicted under the low transport cost assumptions). For substantial increases in the relative price of coal, however, the results are more sensitive to the assumption made about transport costs. In comparing the impacts of the different sulfur-in-fuel standards, therefore, I assume the same transport costs in any comparison.

Costs of Compliance

Table 6-3 shows the costs of compliance with the different sulfur-in-fuel standards, on the assumption of 1969 relative fuel prices and of flue gas desulfurization costs estimated at that time. More recent FGD cost estimates might imply as much as a tripling of these costs, so that the absolute levels in Table 6-3 are probably too low.[1]

These figures imply that a 0.7 percent sulfur-in-fuel standard leads to disproportionately higher compliance costs than a 2.0 percent standard. A 0.7 percent standard implies roughly one-third the sulfur emissions from electric utilities associated with a 2.0 percent standard, but the costs are more than 3 times as great. To be precise, Table 6-3 shows that the costs of the more stringent standard are roughly 5 times those of the less stringent one.

---

[1]The costs shown are those for states burning Appalachian coal, not national costs. In this sense, too, therefore, they are more indicative of the relative costs imposed by a 2.0 and a 0.7 percent standard than of absolute costs.

Table 6-3

COSTS OF COMPLIANCE IN 1980
WITH DIFFERENT SULFUR-IN-FUEL STANDARDS

(Millions of Dollars)

Standard	Low Trans- port Costs	High Trans- port Costs
No Standard	0.0	0.0
2.0 Percent Standard	519.1	550.5
0.7 Percent Standard	2427.3	2604.1

NOTE:

The methods and assumptions underlying these calculations
are discussed in the text.

The increasing costs of sulfur emission restrictions result
mainly from the distribution of sulfur content of Appalachian coals.
Most of them can be washed at relatively low cost to meet a 2.0 percent
standard. To meet a 0.7 percent standard, however, flue gas desulfurization, at much higher cost, is required.[1] In addition, a 0.7 percent
standard requires desulfurization of residual fuel oil that a 2.0
percent standard does not.

I have not attempted to quantify the benefits in air quality,
health, and the quality of life that a 0.7 percent standard would
confer relative to a 2.0 percent standard. The additional costs
seem high enough, however, that a careful analysis of these benefits
should be made to determine if the stricter standard is justified.

Coal Output and Employment in Appalachia

The impact of the different sulfur-in-fuel standards on
Appalachian coal output and employment are shown in Table 6-4.
These figures show that, without the OPEC oil price increases,
Appalachian coal would have lost some markets even in the absence
of sulfur emission restrictions.

---

[1]There is, of course, some Appalachian coal that can meet a
0.7 percent sulfur standard. Much of it is used by steel companies
for coke. Mining costs appear to be much higher, and the supply
curve relatively inelastic for low sulfur coal. See, for example,
the studies surveyed by Richard L. Gordon, *Economic Analysis of
Coal Supply: An Assessment of Existing Studies* (Palo Alto, CA:
Electric Power Research Institute, May 1975).

Table 6-4

APPALACHIAN OUTPUT AND EMPLOYMENT IN 1980
UNDER THE DIFFERENT SULFUR-IN-FUEL STANDARDS,
1969 RELATIVE COAL AND OIL PRICES

	Low Transport Costs		High Transport Costs	
Policy	Total Coal Output	Total Employ-ment	Total Coal Output	Total Employ-ment
Extrapolation	521.3	65235	521.3	65235
No Standard	471.7	59094	503.4	63053
2.0 Percent Standard	530.5	66016	593.4	73913
0.7 Percent Standard	379.5	47875	475.1	59484

UNITS:

Coal output -- Millions of tons

Employment -- Man-years

A 2.0 percent standard puts Appalachian coal in a favorable position relative to midwestern coal, so that such a standard implies an increase in Appalachian coal production even above the level implied by maintaining its share of electric utilities' fossil fuel consumption. Without the additional demand from the Midwest, however, Appalachian coal output would be about 100 million tons less than that shown, or between 20 and 50 million tons below the level without a sulfur-in-fuel standard.[1]

A 0.7 percent standard has a noticeable impact on Appalachian coal output and employment. The impact is more severe if oil transport costs are low -- output is almost 100 million tons less than if no sulfur standard were in effect. This reduction is almost 20 percent of Appalachian coal output in 1980 without a sulfur standard.

If residual fuel oil transport costs are high, however, the impact is much less severe. Indeed, output under a 0.7 percent standard is almost 95 percent of that under no sulfur-in-fuel standard. Given the various errors and uncertainties in the forecasting procedures, a 5 percent decrease is hardly significant.

---

[1]The range arises from the difference in transport costs. With low residual fuel oil transport costs, output would be 50 million tons lower; with high transport costs, output would be only 20 million tons lower.

In summary, it appears that, had relative coal and oil prices remained at their 1969 level, a 0.7 percent sulfur-in-fuel standard would have caused a sharp drop in Appalachian coal output and employment. A 2.0 percent standard, on the other hand, would have caused a sharp increase. The OPEC oil price increases of 1973-1974 resulted in quite a different outcome from either of these, however. In the next chapter, I discuss the impact of OPEC's actions.

Appendix 6A

This appendix contains the supporting detail for the summaries

contained in Table 6-4. For example, Table 6A-1 shows the forecast

of 1980 coal use by electric utilities on the assumption that relative oil

and coal prices stay at 1975 levels. Table 6A-2 shows 1980 Appalachian

production and employment resulting from that forecast. Table 6A-3

shows the costs of stack gas emissions control, by state, under a 0.7

percent sulfur-in-fuel standard. Tables 6A-4 to 6A-6 show the same

information for the scenarios involving 1975 relative fuel prices and

a 2.0 percent sulfur-in-fuel standard.

Tables 6A-7 and 6A-8 show forecasts of utility coal consumption by

state and Appalachian output and employment, for the "status quo"

forecasts (1969 relative fuel prices) and the assumptions called

"low transport costs", while Tables 6A-9 and 6A-10 show the same

information for the "high transport cost" assumptions.

Tables 6A-11 and 6A-12 show the same information for the 0.7 percent

sulfur-in-fuel standard, 1969 relative fuel prices, and low transport

costs, while Table 6A-13 shows the costs of stack gas control under

this scenario. Tables 6A-14 to 6A-16 show the same information under

the same set of assumptions, except for transport costs, which are high.

Tables 6A-17 to 6A-19 show this information for the scenario

involving a 2.0 percent sulfur-in-fuel standard, 1969 relative fuel

prices, and low transport costs, while Tables 6A-20 to 6A-22 are the

same scenario, except for high transport costs.

Table 6A-1

FORECAST OF 1980 COAL USE BY ELECTRIC UTILITIES,
ASSUMING RELATIVE OIL AND COAL PRICES STAY AT 1975 LEVELS

(Millions of Tons)

State, by Region	1980 Forecast
Connecticut	2.33
Massachusetts	2.49
New Hampshire	1.17
New England Total	5.99
New Jersey	3.80
New York	14.73
Pennsylvania	41.72
Middle Atlantic Total	60.25
Illinois	60.46
Indiana	42.12
Michigan	35.31
Ohio	60.58
Wisconsin	14.78
East North Central Total	213.26
Delaware	2.24
District of Columbia	0.48
Florida	8.11
Georgia	13.31
Maryland	11.87
North Carolina	20.24
South Carolina	4.13
Virginia	8.48
West Virginia	41.50
South Atlantic Total	110.36
Alabama	22.58
Kentucky	28.04
Mississippi	0.33
Tennessee	15.65
East South Central Total	66.59
GRAND TOTAL	456.46

Table 6A-2

SCENARIO ASSUMING RELATIVE OIL AND COAL PRICES
STAY AT 1975 LEVELS, SULFUR-IN-FUEL STANDARD OF 0.7 PERCENT

1980 Production and Employment in the
Appalachian Coal Industry, by District

District	(1) Other	(2) Utility	(3) Total	(4) Proportion of 1970 Output	(5) Employment (Man-Years)	(6) Employment as Fraction of Base Case Employment
1	16.53	48.25	64.78	1.389	7014	1.457
2	31.87	11.64	43.51	1.099	6052	1.064
3 & 6	20.81	55.16	75.97	1.518	10606	1.208
4	12.33	67.71	80.04	1.437	8371	1.006
7	42.40	1.48	43.88	1.182	6100	1.018
8	105.90	82.83	188.73	1.172	24421	1.067
13	8.39	16.00	24.39	1.189	2671	1.018
TOTAL	238.23	283.07	521.30	1.270	65235	1.104

NOTE: Base case employment used in column (6) is based on low transport costs.

175

Table 6A-3

COSTS OF STACK GAS EMISSIONS CONTROL, 0.7 PERCENT SULFUR-IN-FUEL STANDARDS, 1975 RELATIVE OIL AND COAL PRICES, ALL OIL IMPORT POLICIES

(Millions of Dollars)

State, by Region	Costs, Plants Using Or Planned to Use Oil	Coal Costs	Total
Connecticut	25.18	11.65	36.83
Maine	0.14	--	0.14
Massachusetts	41.87	12.45	54.32
New Hampshire	1.89	5.85	7.74
Rhode Island	2.90	--	2.90
New England Total	71.98	29.95	101.93
New Jersey	24.56	19.00	43.56
New York	56.73	73.65	130.38
Pennsylvania	28.25	208.60	236.85
Middle Atlantic Total	109.54	301.25	410.79
Illinois	0.41	302.30	302.71
Indiana	0.15	210.60	210.75
Michigan	1.05	176.55	177.60
Ohio	0.32	302.90	303.22
Wisconsin	0.23	73.90	74.13
East North Central Total	2.16	1066.30	1068.46
Delaware	4.75	11.20	15.95
District of Columbia	4.81	2.40	7.21
Florida	46.02	40.55	86.57
Georgia	1.31	66.55	67.86
Maryland	10.46	59.35	69.81
North Carolina	0.04	101.20	101.24
South Carolina	0.36	20.65	21.01
Virginia	7.02	42.40	49.42
West Virginia	0.05	207.50	207.55
South Atlantic Total	74.82	551.80	626.62
Alabama	11.27	112.90	124.17
Kentucky	7.19	140.20	147.39
Mississippi	75.71	1.65	77.36
Tennessee	9.72	78.25	87.97
East South Central Total	103.89	332.95	436.84
Grand Total	362.40	2282.30	2644.70

NOTE: Detail may not add to totals due to independent rounding.

Table 6A-4

ELECTRIC UTILITY COAL CONSUMPTION, 1980,
BY STATES BURNING APPALACHIAN COAL

(Millions of Tons)

State, by Region	Coal Burned	Appalachian Coal[1]	Midwest Coal[2]
Connecticut	2.33		
Massachusetts	2.49		
New Hampshire	1.17		
New England Total	5.99		
New Jersey	3.80		
New York	14.73		
Pennsylvania	41.72		
Middle Atlantic Total	60.25		
Illinois	60.46	53.43	7.03
Indiana	42.12	39.76	2.36
Michigan	35.31		
Ohio	60.58		
Wisconsin	14.78	14.78	0.00
East North Central Total	213.26	107.97	9.39
Delaware	2.24		
District of Columbia	0.48		
Florida	8.11		
Georgia	13.31		
Maryland	11.87		
North Carolina	20.24		
South Carolina	4.13		
Virginia	8.48		
West Virginia	41.50		
South Atlantic Total	110.36		
Alabama	22.58		
Kentucky	28.04		
Mississippi	0.33		
Tennessee	15.65		
East South Central Total	66.59		
GRAND TOTAL	456.46		

[1]Amount of Appalachian coal estimated by model to displace coal from existing sources.

[2]Coal from existing sources estimated to continue to be burned.

177

Table 6A-5

1980 PRODUCTION AND EMPLOYMENT
IN THE APPALACHIAN COAL INDUSTRY, BY DISTRICT

	(1)	(2)	(3)	(4)	(5)	(6)
				Proportion of 1970 Output	Employment (Man-Years)	Employment as Fraction of Base Case Employment
District	Other	Utility	Total			
1	16.53	48.25	64.78	1.389	7014	1.457
2	31.87	11.64	43.51	1.099	6052	1.064
3 & 6	20.81	76.75	97.56	1.949	13620	1.552
4	12.33	110.51	122.84	2.205	12847	1.544
7	42.40	1.48	43.88	1.182	6100	1.018
8	105.90	123.20	229.10	1.423	29645	1.296
13	8.39	16.00	24.39	1.189	2671	1.018
TOTAL	238.23	387.83	626.06	1.525	77949	1.327

NOTE: Base case employment used in column (6) is based on low transport costs.

Table 6A-6

COSTS OF STACK GAS EMISSIONS CONTROL, 2.0 PERCENT
SULFUR-IN-FUEL STANDARD, 1975 RELATIVE OIL AND COAL PRICES[1]

State, by Region	Coal Costs
Connecticut	1.75
Maine	--
Massachusetts	1.87
New Hampshire	0.88
Rhode Island	--
New England Total	4.49
New Jersey	2.85
New York	11.05
Pennsylvania	31.29
Middle Atlantic Total	45.19
Illinois	169.98
Indiana	119.14
Michigan	26.49
Ohio	45.43
Wisconsin	24.17
East North Central Total	385.21
Delaware	1.68
District of Columbia	0.36
Florida	6.08
Georgia	9.98
Maryland	8.90
North Carolina	15.18
South Carolina	3.10
Virginia	6.36
West Virginia	31.13
South Atlantic Total	82.77
Alabama	16.94
Kentucky	21.03
Mississippi	0.25
Tennessee	11.74
East South Central Total	49.94
GRAND TOTAL	567.60

[1]Additional costs of desulfurizing residual
to meet a 2.0 percent sulfur-in-fuel standard
are assumed to be zero.

179

Table 6A-7

ELECTRIC UTILITY COAL CONSUMPTION, 1980,
BY STATES BURNING APPALACHIAN COAL
STATUS QUO FORECASTS, LOW TRANSPORT COSTS

(Millions of Tons)

State, by Region	Coal Burned
Connecticut	0.00
Massachusetts	0.00
New Hampshire	0.00
New England Total	0.00
New Jersey	0.00
New York	9.18
Pennsylvania	33.01
Middle Atlantic Total	42.20
Illinois	60.46
Indiana	42.12
Michigan	35.31
Ohio	60.18
Wisconsin	14.63
East North Central Total	212.71
Delaware	1.19
District of Columbia	0.48
Florida	1.02
Georgia	12.92
Maryland	0.37
North Carolina	18.08
South Carolina	3.48
Virginia	3.22
West Virginia	40.71
South Atlantic Total	81.48
Alabama	22.58
Kentucky	28.04
Mississippi	0.33
Tennessee	15.65
East South Central Total	66.59
GRAND TOTAL	402.98

Table 6A-8

1980 PRODUCTION AND EMPLOYMENT IN THE APPALACHIAN COAL INDUSTRY,
BY DISTRICT
STATUS QUO FORECASTS, LOW TRANSPORT COSTS

District	(1) Other	(2) Utility	(3) Total	(4) Proportion of 1970 Output	(5) Employment (Man-Years)
1	16.53	27.92	44.45	0.953	4813
2	31.87	9.03	40.90	1.033	5689
3 & 6	20.81	42.06	62.87	1.256	8777
4	12.33	67.24	79.56	1.428	8320
7	42.40	0.72	43.12	1.162	5994
8	105.90	70.90	176.80	1.098	22878
13	8.39	15.58	23.96	1.168	2624
TOTAL	238.23	233.45	471.68	1.149	59094

Table 6A-9

FORECASTS OF 1980 ELECTRIC UTILITY COAL CONSUMPTION,
BY STATES BURNING APPALACHIAN COAL
STATUS QUO FORECASTS, HIGH TRANSPORT COSTS

(Millions of Tons)

State, by Region	Coal Burned
Connecticut	0.93
Massachusetts	0.00
New Hampshire	0.00
New England Total	0.93
New Jersey	3.00
New York	14.73
Pennsylvania	35.98
Middle Atlantic Total	53.71
Illinois	60.46
Indiana	42.12
Michigan	35.31
Ohio	60.58
Wisconsin	14.78
East North Central Total	213.26
Delaware	2.24
District of Columbia	0.48
Florida	7.62
Georgia	13.31
Maryland	8.18
North Carolina	20.24
South Carolina	4.08
Virginia	6.17
West Virginia	41.50
South Atlantic Total	103.83
Alabama	22.58
Kentucky	28.04
Mississippi	0.33
Tennessee	15.65
East South Central Total	66.59
GRAND TOTAL	438.32

Table 6A-10

1980 PRODUCTION AND EMPLOYMENT IN THE APPALACHIAN COAL INDUSTRY,
BY DISTRICT
STATUS QUO FORECASTS, HIGH TRANSPORT COSTS

District	(1) Other	(2) Utility	(3) Total	(4) Proportion of 1970 Output	(5) Employment (Man-Years)
1	16.53	39.09	55.62	1.192	6022
2	31.87	10.52	42.39	1.071	5896
3 & 6	20.81	50.50	71.32	1.425	9957
4	12.33	67.71	80.04	1.437	8371
7	42.40	1.22	43.63	1.175	6065
8	105.90	80.16	186.05	1.155	24075
13	8.39	15.97	24.36	1.188	2668
TOTAL	238.23	265.18	503.40	1.226	63053

183

Table 6A-11
ELECTRIC UTILITY COAL CONSUMPTION, 1980, BY STATES BURNING APPALACHIAN COAL 0.7 PERCENT SULFUR-IN-FUEL STANDARD, LOW TRANSPORT COSTS AND PRICES
(Millions of Tons)

State, by Region	Coal Burned
Connecticut	0.00
Massachusetts	0.00
New Hampshire	0.00
New England Total	0.00
New Jersey	0.00
New York	5.65
Pennsylvania	12.90
Middle Atlantic Total	18.55
Illinois	59.93
Indiana	40.96
Michigan	35.31
Ohio	34.77
Wisconsin	14.48
East North Central Total	185.45
Delaware	1.19
District of Columbia	0.00
Florida	0.00
Georgia	0.00
Maryland	0.00
North Carolina	0.00
South Carolina	0.00
Virginia	1.08
West Virginia	34.61
South Atlantic Total	36.88
Alabama	12.68
Kentucky	25.64
Mississippi	0.33
Tennessee	14.47
East South Central Total	53.11
GRAND TOTAL	293.99

Table 6A-12

1980 PRODUCTION AND EMPLOYMENT IN THE APPALACHIAN COAL INDUSTRY,
BY DISTRICT
0.7 PERCENT SULFUR-IN-FUEL STANDARD, LOW TRANSPORT COSTS AND PRICES

District	(1) Other	(2) Utility	(3) Total	(4) Proportion of 1970 Output	(5) Employment (Man-Years)
1	16.53	14.39	30.92	0.663	3348
2	31.87	4.22	36.09	0.912	5020
3 & 6	20.81	29.71	50.52	1.009	7053
4	12.33	48.96	61.29	1.100	6410
7	42.40	0.20	42.60	1.147	5922
8	105.90	35.31	141.21	0.877	18272
13	8.39	8.52	16.90	0.824	1851
TOTAL	238.23	141.31	379.53	0.924	47875

Table 6A-13

COSTS OF STACK GAS EMISSIONS CONTROL, 0.7 PERCENT
SULFUR-IN-FUEL STANDARD, 1980, LOW OIL TRANSPORT COSTS

(Millions of Dollars)

State, by Region	Costs, Plants Using Or Planned to Use Oil	Coal Costs	Costs at Converted Plants	Total
Connecticut	25.18	0.00	6.79	31.97
Maine	0.14	-	-	0.14
Massachusetts	41.87	0.00	7.41	49.28
New Hampshire	1.89	0.00	3.87	5.76
Rhode Island	2.90	-	-	2.90
New England Total	71.98	0.00	18.06	90.04
New Jersey	24.56	0.00	11.86	36.42
New York	56.73	28.27	31.02	116.02
Pennsylvania	28.25	64.49	111.85	204.59
Middle Atlantic Total	109.54	92.76	154.72	357.03
Illinois	0.41	299.67	2.18	302.26
Indiana	0.15	204.80	4.85	209.80
Michigan	1.05	176.57	0.00	177.62
Ohio	0.32	173.84	90.92	265.08
Wisconsin	0.23	72.39	0.89	73.51
East North Central Total	2.16	927.27	98.84	1028.27
Delaware	4.75	5.97	3.21	13.93
District of Columbia	4.81	0.00	1.64	6.45
Florida	46.02	0.00	23.12	69.14
Georgia	1.31	0.00	57.13	58.44
Maryland	10.46	0.00	36.75	47.21
North Carolina	0.04	0.00	78.06	78.10
South Carolina	0.36	0.00	13.64	14.00
Virginia	7.02	5.39	26.15	38.56
West Virginia	0.05	173.04	30.00	203.09
South Atlantic Total	74.82	184.40	269.70	528.92
Alabama	11.27	63.38	38.93	113.58
Kentucky	7.19	128.20	10.15	145.54
Mississippi	75.71	1.64	0.00	77.35
Tennessee	9.72	72.34	4.50	86.56
East South Central Total	103.89	265.56	53.58	423.03
GRAND TOTAL	362.40	1469.99	594.90	2427.29

Table 6A-14

ELECTRIC UTILITY COAL CONSUMPTION, 1980,
BY STATES BURNING APPALACHIAN COAL
0.7 PERCENT SULFUR-IN-FUEL STANDARD,
HIGH TRANSPORT COSTS AND PRICES

(Millions of Tons)

State, by Region	Coal Burned
Connecticut	0.00
Massachusetts	0.00
New Hampshire	0.00
New England Total	0.00
New Jersey	0.00
New York	10.18
Pennsylvania	33.01
Middle Atlantic Total	43.19
Illinois	60.46
Indiana	42.12
Michigan	35.31
Ohio	60.31
Wisconsin	14.63
East North Central Total	212.83
Delaware	1.19
District of Columbia	0.07
Florida	4.14
Georgia	12.92
Maryland	5.13
North Carolina	12.86
South Carolina	2.55
Virginia	5.71
West Virginia	40.71
South Atlantic Total	85.29
Alabama	22.58
Kentucky	28.04
Mississippi	0.33
Tennessee	15.65
East South Central Total	66.59
GRAND TOTAL	407.90

Table 6A-15

1980 PRODUCTION AND EMPLOYMENT IN THE APPALACHIAN COAL INDUSTRY,
BY DISTRICT
0.7 PERCENT SULFUR-IN-FUEL STANDARD, HIGH TRANSPORT COSTS AND PRICES

District	(1) Other	(2) Utility	(3) Total	(4) Proportion of 1970 Output	(5) Employment (Man-Years)
1	16.53	31.09	47.62	1.021	5156
2	31.87	9.20	41.06	1.037	5711
3 & 6	20.81	44.22	65.03	1.299	9079
4	12.33	67.35	79.67	1.430	8332
7	42.40	0.81	43.22	1.164	6008
8	105.90	68.42	174.31	1.083	22555
13	8.39	15.75	24.14	1.177	2644
TOTAL	238.23	236.84	475.06	1.157	59484

Table 6A-16

COSTS OF STACK GAS EMISSIONS CONTROL, 0.7 PERCENT
SULFUR-IN-FUEL STANDARD, 1980, HIGH TRANSPORT COSTS
(Millions of Dollars)

State, by Region	Costs, Plants Using, Or Planned to Use Oil	Coal Costs	Costs at Converted Plants	Total
Connecticut	25.18	0.00	9.05	34.23
Maine	0.14	–	–	0.14
Massachusetts	41.87	0.00	9.88	51.75
New Hampshire	1.89	0.00	5.16	7.05
Rhode Island	2.90	–	–	2.90
New England Total	71.98	0.00	24.09	96.07
New Jersey	24.56	0.00	15.91	40.47
New York	56.73	50.89	18.97	126.59
Pennsylvania	28.25	165.07	33.88	227.20
Middle Atlantic Total	109.54	215.96	68.76	394.26
Illinois	0.41	302.32	0.00	302.73
Indiana	0.15	210.61	0.00	210.76
Michigan	1.05	176.57	0.00	177.62
Ohio	0.32	301.53	1.21	303.06
Wisconsin	0.23	73.14	0.67	74.04
East North Central Total	2.16	1064.17	1.88	1068.21
Delaware	4.75	5.97	4.28	15.00
District of Columbia	4.81	0.36	1.82	6.99
Florida	46.02	20.70	15.24	81.96
Georgia	1.31	64.60	1.54	67.45
Maryland	10.46	25.66	27.89	64.01
North Carolina	0.04	64.30	35.00	99.34
South Carolina	0.36	12.74	7.11	20.21
Virginia	7.02	28.56	11.43	47.01
West Virginia	0.05	203.56	3.14	206.75
South Atlantic Total	74.82	426.45	107.47	608.74
Alabama	11.27	112.89	0.00	124.16
Kentucky	7.19	140.18	0.00	147.37
Mississippi	75.71	1.64	0.00	77.35
Tennessee	9.72	78.25	0.00	87.97
East South Central Total	103.89	332.96	0.00	436.85
GRAND TOTAL	362.40	2039.54	202.19	2604.13

Table 6A-17

ELECTRIC UTILITY COAL CONSUMPTION, 1980,
BY STATES BURNING APPALACHIAN COAL
2.0 PERCENT SULFUR-IN-FUEL STANDARD, LOW TRANSPORT COSTS AND PRICES

(Millions of Tons)

State, by Region	Coal Burned	Appalachian Coal[1]	Midwest Coal[2]
Connecticut	0.00		
Massachusetts	0.00		
New Hampshire	0.00		
New England Total	0.00		
New Jersey	0.00		
New York	8.13		
Pennsylvania	32.55		
Middle Atlantic Total	40.68		
Illinois	60.46	53.43	7.03
Indiana	42.12	39.76	2.36
Michigan	35.31		
Ohio	53.56		
Wisconsin	14.48	14.48	0.00
East North Central Total	205.93	107.67	9.39
Delaware	1.19		
District of Columbia	0.07		
Florida	1.02		
Georgia	10.61		
Maryland	0.00		
North Carolina	11.35		
South Carolina	0.96		
Virginia	3.22		
West Virginia	40.71		
South Atlantic Total	69.14		
Alabama	19.64		
Kentucky	28.04		
Mississippi	0.33		
Tennessee	15.65		
East South Central Total	63.65		
GRAND TOTAL	379.40		

[1]Amount of Appalachian coal estimated by model to
displace coal from existing sources.

[2]Coal from existing sources estimated to continue
to be burned.

190

Table 6A-18

1980 PRODUCTION AND EMPLOYMENT IN THE APPALACHIAN COAL INDUSTRY,
BY DISTRICT

2.0 PERCENT SULFUR-IN-FUEL STANDARD, LOW TRANSPORT COSTS

District	(1) Other	(2) Utility	(3) Total	(4) Proportion of 1970 Output	(5) Employment (Man-Years)
1	16.53	26.68	43.22	0.926	4680
2	31.87	8.62	40.49	1.023	5632
3 & 6	20.81	61.91	80.72	1.613	11269
4	12.33	105.07	117.40	2.108	12278
7	42.40	0.54	42.94	1.157	5969
8	105.90	97.89	183.79	1.141	23782
13	8.39	13.59	21.98	1.071	2407
TOTAL	238.23	314.30	530.54	1.292	66016

Table 6A-19

COSTS OF STACK GAS EMISSIONS CONTROL, 2.0 PERCENT
SULFUR-IN-FUEL STANDARD, 1980, LOW TRANSPORT COSTS

(Millions of Dollars)

State, by Region	Costs, Plants Using, Or Planned to Use Oil	Coal Costs	Costs at Converted Plants	Total
Connecticut	0.00	0.00	0.00	0.00
Maine	0.00	-	-	-
Massachusetts	0.00	0.00	0.00	0.00
New Hampshire	0.00	0.00	0.00	0.00
Rhode Island	0.00	-	-	-
New England Total	0.00	0.00	0.00	0.00
New Jersey	0.00	0.00	0.00	0.00
New York	0.00	6.10	0.34	6.44
Pennsylvania	0.00	24.41	0.33	24.74
Middle Atlantic Total	0.00	30.51	0.67	31.18
Illinois	0.00	169.98	0.00	169.98
Indiana	0.00	119.14	0.00	119.14
Michigan	0.00	26.49	0.00	26.49
Ohio	0.00	40.17	3.61	43.78
Wisconsin	0.00	23.69	0.03	23.72
East North Central Total	0.00	379.47	3.64	383.11
Delaware	0.00	0.90	0.00	0.90
District of Columbia	0.00	0.05	0.01	0.06
Florida	0.00	0.76	0.00	0.76
Georgia	0.00	7.96	1.18	9.14
Maryland	0.00	0.00	0.00	0.00
North Carolina	0.00	8.52	2.00	10.52
South Carolina	0.00	0.72	0.45	1.17
Virginia	0.00	2.42	0.00	2.42
West Virginia	0.00	30.53	0.00	30.53
South Atlantic Total	0.00	51.86	3.64	55.50
Alabama	0.00	14.73	1.55	16.28
Kentucky	0.00	21.03	0.00	21.03
Mississippi	0.00	0.25	0.00	0.25
Tennessee	0.00	11.74	0.00	11.74
East South Central Total	0.00	47.75	1.55	49.30
GRAND TOTAL	0.00	509.59	9.50	519.09

Table 6A-20

ELECTRIC UTILITY COAL CONSUMPTION, 1980,
BY STATES BURNING APPALACHIAN COAL
2.0 PERCENT SULFUR-IN-FUEL STANDARD, HIGH TRANSPORT COSTS AND PRICES

(Millions of Tons)

State, by Region	Coal Burned	Appalachian Coal[1]	Midwest Coal[2]
Connecticut	0.00		
Massachusetts	0.00		
New Hampshire	0.00		
New England Total	0.00		
New Jersey	0.11		
New York	11.57		
Pennsylvania	35.31		
Middle Atlantic Total	46.99		
Illinois	60.46	53.43	7.03
Indiana	42.12	39.76	2.36
Michigan	35.31		
Ohio	60.58		
Wisconsin	14.64	14.64	0.00
East North Central Total	213.11	107.83	9.39
Delaware	1.19		
District of Columbia	0.48		
Florida	6.59		
Georgia	12.92		
Maryland	6.80		
North Carolina	19.27		
South Carolina	4.08		
Virginia	6.17		
West Virginia	41.50		
South Atlantic Total	99.01		
Alabama	22.58		
Kentucky	28.04		
Mississippi	0.33		
Tennessee	15.65		
East South Central Total	66.59		
GRAND TOTAL	425.71		

[1]Amount of Appalachian coal estimated by model to displace coal from existing sources.

[2]Coal from existing sources estimated to continue to be burned.

193

Table 6A-21

1980 PRODUCTION AND EMPLOYMENT IN THE APPALACHIAN COAL INDUSTRY,
BY DISTRICT

2.0 PERCENT SULFUR-IN-FUEL STANDARD, HIGH TRANSPORT COSTS AND PRICES

District	(1) Other	(2) Utility	(3) Total	(4) Proportion of 1970 Output	(5) Employment (Man-Years)
1	16.53	34.16	50.69	1.087	5488
2	31.87	9.81	41.68	1.053	5797
3 & 6	20.81	67.66	88.47	1.767	12351
4	12.33	110.16	122.49	2.199	12810
7	42.40	1.05	43.45	1.170	6040
8	105.90	116.42	222.32	1.381	28768
13	8.39	15.89	24.28	1.184	2659
TOTAL	238.23	355.15	593.38	1.445	73913

194

Table 6A-22

COSTS OF STACK GAS EMISSIONS CONTROL, 2.0 PERCENT
SULFUR-IN-FUEL STANDARD, 1980, HIGH TRANSPORT COSTS

(Millions of Dollars)

State, by Region	Costs, Plants Using, Or Planned to Use Oil	Coal Costs	Costs at Converted Plants	Total
Connecticut	0.00	0.00	0.54	0.54
Maine	0.00	-	-	-
Massachusetts	0.00	0.00	0.00	0.00
New Hampshire	0.00	0.00	0.00	0.00
Rhode Island	0.00	-	-	-
New England Total	0.00	0.00	0.54	0.54
New Jersey	0.00	0.09	1.13	1.22
New York	0.00	8.68	2.29	10.97
Pennsylvania	0.00	26.48	0.31	26.79
Middle Atlantic Total	0.00	35.25	3.73	38.98
Illinois	0.00	169.98	0.00	169.98
Indiana	0.00	119.14	0.00	119.14
Michigan	0.00	26.49	0.00	26.49
Ohio	0.00	45.43	0.00	45.43
Wisconsin	0.00	23.72	0.34	24.07
East North Central Total	0.00	384.76	0.34	385.10
Delaware	0.00	0.90	0.16	1.06
District of Columbia	0.00	0.36	0.00	0.36
Florida	0.00	4.94	0.67	5.61
Georgia	0.00	9.69	0.27	9.96
Maryland	0.00	5.10	0.11	5.22
North Carolina	0.00	14.45	0.40	14.85
South Carolina	0.00	3.06	0.00	3.06
Virginia	0.00	4.63	0.00	4.63
West Virginia	0.00	31.13	0.00	31.13
South Atlantic Total	0.00	74.26	1.61	75.88
Alabama	0.00	16.93	0.00	16.93
Kentucky	0.00	21.03	0.00	21.03
Mississippi	0.00	0.25	0.00	0.25
Tennessee	0.00	11.74	0.00	11.74
East South Central Total	0.00	49.95	0.00	49.95
GRAND TOTAL	0.00	544.22	6.22	550.45

Chapter 7

OPEC'S PRICE INCREASES

Between 1970 and 1973, the real price of crude oil increased
substantially in percentage terms but modestly in absolute amounts.
These increases largely resulted from increases in taxes and royalties
levied by host country governments. Between September 1973 and May
1974, however, the Organization of Petroleum Exporting Countries
(OPEC) discovered and exploited its ability to increase taxes and
royalties to levels much higher than previously observed.

Because residual fuel oil is the main fossil fuel substitute
for coal, and because the price of fuel oil follows the price of
crude oil, OPEC's actions affected the market for coal as well as
that for petroleum products. In this chapter I consider the impact
of the cartel's actions on the demand for Appalachian coal.

After a brief review of the actions and their impact on fuel
oil and coal prices, I present a forecast of Appalachian coal
consumption in 1980. This forecast is naive in that it is not

explicitly based on the economics of converting existing oil-burning plants to coal, but is nevertheless useful because estimates of conversion costs (especially the additional costs of sulfur emission controls) are at best rough guesses. The method thus provides another set of estimates to those in the next section, where I use existing estimates of conversion costs to forecast the coal that will be consumed by plants converting from oil, leading to another set of forecasts of 1980 demand for Appalachian coal. Finally, I compare the pre-embargo and post-embargo forecasts to assess the impact of the OPEC price increases.

The Oil Price Increases

Although world oil prices started to increase by 1970, the increases up to September 1973 were not large in absolute amounts. For example, the market price (unadjusted for inflation) of Libyan oil rose from about $1.60 per barrel in the summer of 1970 to $3.35 per barrel by the summer of 1973.[1] After the Yom Kippur war in October 1973, however, production cutbacks and price increases by OPEC members caused world oil prices to increase sharply. Table 7-1 shows that the average price paid by U.S. refiners for imported crude oil more than tripled between May 1973 and May 1974. Since then, prices have increased somewhat, and OPEC will no doubt attempt to

---

[1]*New York Times,* September 23, 1970, p. 65:7 and *Oil and Gas Journal,* August 27, 1973, p. 30. These prices are those at Libyan terminals and do not include transportation costs and duties to the United States.

Table 7-1

RECENT IMPORTED CRUDE OIL PRICES

Year	Month	Weighted Average Crude Oil Cost, Imported Oil, Dollars per Barrel
1973	May	3.92
	November	6.49
	December	8.22
1974	January	9.59
	February	12.45
	March	12.73
	April	12.72
	May	13.02
	June	13.06
	July	12.75
	August	12.68
	September	12.53
	October	12.44
	November	12.53
	December	12.82
1975	January	12.77
	February	13.05
	March	13.28
	April	13.26
	May	13.27
	June	14.15
	July	14.03
	August	14.25
	September	14.04
	October	14.66
	November	15.04
	December	14.81
1976	January	13.27

SOURCE: Federal Energy Administration, *Month Energy Review,* June 1976, p. 65.

198

continue to raise prices as long as it can profitably do so.[1]

The price of residual fuel oil has generally kept pace with the price of crude oil. The average spot price of residual bought by utilities along the eastern seaboard in December 1974 ranged from $9.50 to $13.00 per barrel.[2] These prices are roughly five times their 1970 level.[3] As can be seen in Table 7-2, which shows average as-burned costs by state for 1974, even the considerable increase in coal prices since 1969 has been much less than the increase in oil prices.[4]

---

[1]This statement is intentionally vague, because I am not aware of any studies that have determined the profit-maximizing price. That price is almost certainly time-dependent, as demand becomes more elastic over time and as new supplies come on to the market. OPEC must also resolve the problem of apportioning reductions in output among its members and of maintaining a uniform price policy. Its continued success is a matter for speculation.

[2]Some very small quantities were bought at slightly higher prices. Federal Power Commission, *FPC News,* Vol. 8, No. 13 (March 28, 1975), p. 20.

[3]Direct comparison with Table 3-3 is somewhat obscured by differences in sulfur content, but the general magnitude is reasonably clear. For example, average prices paid in December 1974 for fuel oil having between 2.01 and 3 percent sulfur (probably most similar to the oil for which prices are shown in Table 3-3, except for Boston, New York, and Philadelphia, which were burning low sulfur oil) ranged between $10.41 and $11.60 per barrel (*ibid.*).

[4]South Carolina is the only state shown in which coal costs exceed oil costs. This anomaly is probably due to the price of residual fuel oil being below current market levels. All of the residual fuel oil used by South Carolina utilities was bought under contract, at a price around $7.00 per barrel (*ibid.*). If this oil were valued at market clearing prices, its price would be around $1.80 per million BTU, much more expensive than coal.

Table 7-2

PATTERNS OF UTILITY FUEL CONSUMPTION AND AS-BURNED PRICES,
BY STATE, 1974

State, by Region	% of Total BTU			Average Price, ¢/MMBTU		
	Coal	Oil	Gas	Coal	Oil	Gas
Connecticut	2.5	97.5	0.0	110.5	205.9	–
Massachusetts	12.6	85.2	2.2	134.0	194.8	144.2
New Hampshire	79.1	20.9	0.0	81.0	184.4	–
New England Total	13.1	85.4	1.5	114.5	196.7	128.6
New Jersey	26.7	69.2	4.2	134.4	213.9	70.7
New York	21.1	75.0	3.9	106.1	202.7	68.8
Pennsylvania	87.4	12.4	0.2	78.4	206.3	154.0
Middle Atlantic Total	54.3	43.6	2.1	86.1	205.6	73.4
Illinois	88.5	6.3	5.1	54.6	139.6	76.3
Indiana	95.7	2.4	1.9	46.3	174.3	56.0
Michigan	80.7	13.1	6.2	82.9	182.3	98.9
Ohio	97.3	1.6	1.1	87.6	222.9	69.4
Wisconsin	89.6	0.6	9.8	72.6	167.3	56.3
East North Central Total	91.3	4.9	3.8	70.1	173.2	77.0
Delaware	29.7	69.9	0.4	110.4	224.6	181.1
District of Columbia	25.6	74.4	0.0	168.4	199.9	–
Florida	21.0	57.3	21.7	73.8	179.9	58.6
Georgia	76.7	9.4	13.8	78.9	168.9	58.3
Maryland	38.2	61.8	0.0	120.9	184.3	–
North Carolina	95.4	4.5	0.2	107.0	180.1	91.9
South Carolina	76.0	21.1	3.0	130.4	114.1	75.8
Virginia	40.3	59.4	0.3	121.5	165.1	55.2
West Virginia	99.8	0.1	0.0	85.3	225.8	27.4
South Atlantic Total	62.8	30.3	6.9	97.3	177.9	59.2
Alabama	98.5	0.1	1.4	74.8	203.8	85.3
Kentucky	98.8	0.0	1.2	54.9	221.0	40.5
Mississippi	28.1	43.4	28.5	60.6	179.1	55.8
Tennessee	99.9	0.1	0.0	49.6	265.6	–
East South Central Total	92.5	4.0	3.4	59.5	179.9	57.2
GRAND TOTAL	71.6	24.2	4.2	78.0	191.2	66.9

SOURCE: Federal Power Commission, *FPC News*, Vol. 8, No. 13 (March 28, 1975), p. 30.

These figures, which are a mixture of spot and contract prices, are the only ones readily available for the entire year. The contract prices for a single month, December 1974, are shown in Table 7-3. Although, for any particular state, average December 1974 contract prices may differ from average 1974 prices for both coal and oil, the differences do not appear to be systematic. Use of contract prices instead of average prices would not lead to different conclusions about the size of the oil price increases or the gap between relative oil and coal prices.

The disparity in prices brought about by OPEC's actions may be seen more clearly in Table 7-4, which shows the difference between 1974 average oil and coal costs, by state. With the exception of South Carolina (see footnote 4 on p. 199), the average cost of oil ranges between $0.30 and $2.16 per million BTU above the cost of coal, with the overall average being $1.13. The implications of these price increases for consumption of Appalachian coal are discussed in the rest of this chapter.

Impact on Coal Consumption: An Extrapolation

The sharp increase in oil prices has two conflicting impacts on the demand for coal. On the one hand, since coal is a substitute for residual fuel oil, the demand for coal increases. On the other, since fuel costs account for a large proportion of the costs of generating electricity and since, sooner or later, these costs are reflected in electric rates, consumption of electricity tends to

201

Table 7-3
DECEMBER 1974 CONTRACT FUEL OIL AND COAL PRICES
(¢/MMBTU)

State, by Region	(1) Coal	(2) Residual Fuel Oil	(3) Difference (2) - (1)
Connecticut	–	210.1	117.5[1]
Massachusetts	91.6	209.1	117.5
New Hampshire	93.9	188.1	94.2
New England Total	93.5	209.1	115.6
New Jersey	186.1	216.3	30.2
New York	108.6	209.0	100.4
Pennsylvania	105.3	231.4	126.1
Middle Atlantic Total	112.2	213.0	100.8
Illinois	66.5	127.9	61.9
Indiana	49.4	–	61.9[2]
Michigan	91.8	171.9	80.1
Ohio	82.2	224.8	142.6
Wisconsin	81.5	182.8	101.3
East North Central Total	70.8	160.4	89.6
Delaware	96.7	197.1	100.4
District of Columbia	166.0	208.1	42.1
Florida	88.4	188.4	100.0
Georgia	79.2	172.3	93.1
Maryland	165.4	205.8	40.4
North Carolina	97.1	189.9	92.8
South Carolina	137.1	117.1	-20.0
Virginia	112.3	178.6	66.3
West Virginia	99.3	–	66.3[3]
South Atlantic Total	99.1	189.8	85.7
Alabama	73.7	–	66.5[4]
Kentucky	50.2	–	66.5[4]
Mississippi	70.3	136.8	66.5[4]
Tennessee	65.3	–	66.5[4]
East South Central Total	62.8	136.8	66.5[4]

GRAND TOTAL

SOURCE: *FPC News*, Vol. 8, No. 13, March 28, 1975, pp. 14, 20.

[1]Massachusetts difference used
[2]Illinois difference used
[3]Virginia difference used
[4]Mississippi difference used

Table 7-4

AVERAGE 1974 COST DIFFERENCE BETWEEN OIL AND COAL, BY STATE

(Cents per Million BTU)

State, by Region	Oil Price Minus Coal Price	Average December 1974 Spot Coal Price	Average Oil Price Minus December 1974 Spot Coal Price
Connecticut	95.4	–	–
Massachusetts	60.8	–	–
New Hampshire	103.4	–	–
New England Total	82.2	–	–
New Jersey	79.5	199.3	14.6
New York	96.6	135.9	66.8
Pennsylvania	127.9	109.6	97.7
Middle Atlantic Total	119.5	117.3	88.3
Illinois	85.0	109.6	30.0
Indiana	128.0	91.6	82.7
Michigan	99.4	183.3	-1.0
Ohio	135.3	142.9	80.0
Wisconsin	94.7	84.1	83.2
East North Central Total	103.1		
Delaware	114.2	134.4	90.2
District of Columbia	31.5	219.5	-19.6
Florida	106.1	150.7	29.2
Georgia	90.0	178.0	-9.1
Maryland	63.4	147.8	37.5
North Carolina	73.1	164.9	15.2
South Carolina	-16.3	167.8	-53.7
Virginia	43.6	189.4	-14.3
West Virginia	140.5	151.0	74.8
South Atlantic Total	80.6	163.9	74.0
Alabama	129.0	159.9	43.9
Kentucky	166.1	133.5	87.5
Mississippi	118.5	–	–
Tennessee	216.0	130.8	135.2
East South Central Total	120.4	142.3	37.6
GRAND TOTAL	113.2		

SOURCE: Derived from Table 7-3.

decrease, leading in turn to a reduction in coal consumption. Other factors, chiefly difficulties in power plant siting and increases in construction costs, have also contributed to the sharp rise in electric rates, and as a result, forecasts of electricity consumption made in the late 1960's and early 1970's have proved overly optimistic.

The immediate impact on net electricity generation can be seen in Table 7-5, which compares the growth rate based on experience before 1970 with the actual growth rates between 1969 and 1973 and between 1969 and 1975. As can be seen, the growth in electricity generation between 1969 and 1973 was, on the whole, quite close to that forecasted, although the regional detail was less accurate. By 1975, however, net electricity generation was at roughly its 1973 level, leading to the lower growth rates shown in the third column of Table 7-5.[1]

---

[1]Some very rough calculations may put these changes in better perspective. If we assume that, except for the oil price increases, net electricity generation would have grown between 1973 and 1975 at the same rate it grew between 1969 and 1973, then the reduction attributable to the higher oil prices was about 12.1 percent. This reduction is almost exactly what Joskow and Baughman's estimates would have predicted (between 10 and 11 percent, based on wholesale and consumer price indices for electricity reported in Bureau of Labor Statistics, *Handbook of Labor Statistics 1976*, pp. 250 and 262). (Paul L. Joskow and Martin L. Baughman, "The Future of the U.S. Nuclear Energy Industry", *The Bell Journal of Economics and Management Science*, Spring 1976, pp. 3-32. I derived the 2-year elasticities from the 1-year elasticity and the long-run elasticity by assuming the lag structure could be approximated by a Koyck lag. The elasticities are shown on Table 1, p. 14.) It is higher than the 5 percent implied by Griffin's estimated elasticities (James M. Griffin, "The Effects of Higher Prices on Electricity Consumption", *The Bell Journal of Economics and Management Science*, Autumn 1974, pp. 515-539. I assumed that the 2-year elasticity stood in the same proportion to the 2-year coefficients that the 1-year elasticity did to the 1-year coefficient on price, to obtain the 2-year elasticity. Table 2, p. 520 and Table 3, p. 530).
Inasmuch as the price of oil was roughly 300 percent higher over this period, but the wholesale price index of electricity increased by 49.6 percent and the consumer price index of electricity by 33.7 percent, and consumption was 12 percent below what it would have been, the conclusion that Griffin drew from his simulations appears to apply here: "First, higher fossil-fuel prices in Case B have a relatively small impact on electricity prices and an even smaller effect upon electricity demand" (Griffin, p. 535).

Table 7-5

ACTUAL AND FORECASTED REGIONAL GROWTH RATES
IN NET ELECTRICITY GENERATION, 1969-1975

Region	Forecasted Growth Rate	Growth Rate, 1969-1973	Growth Rate, 1969-1975
New England	6.43	6.68	4.14
Middle Atlantic	6.43	5.94	3.45
East North Central	8.57	5.37	4.04
South Atlantic	6.69	9.17	6.65
East South Central	6.86	5.41	2.66
TOTAL	7.33	6.61	4.46

NOTE: All growth rates on the basis of continuous compounding.

SOURCES: Forecasts -- see Table 4-3.

Actual -- computed from data on net electricity generated by electric utilities from fuel (excludes hydro); 1969 data from Edison Electric Institute, *Statistical Yearbook of the Electric Utility Industry for 1970*, Table 14S; 1973 and 1975 data from *FPC News*, Volume 9, Number 23, June 4, 1976, pp. 18-20.

One way to forecast Appalachian coal consumption under the changed market conditions is to assume that coal's share of electricity generated by fossil fuels remains constant between 1975 and 1980. That is, if, with the shift in relative prices after 1973-1974, the adjustments were made quickly enough that coal's share did not change after 1975, this assumption would be accurate. Along with this assumption, the only other element needed for the forecast is a forecast of net electricity generated by fossil fuels. This latter I derived by subtracting an estimate of electricity generated by nuclear power plants from a forecast of net electricity generated by all fuels (thereby excluding electricity generated by hydroelectric plants). The procedures used to derive these estimates are, briefly, as follows.

Net Electricity Generation

Using the 1975 level as a base, I extrapolated net generation to 1980 using the growth rates implied by the forecasts supplied by the regional reliability councils.[1] These rates are, on average, slightly higher than average growth rates over the 1969 to 1975 period. Using the 1969-1975 actual growth rates instead of the reliability council forecasts leads to a forecast of 1980 net generation that is about 6 percent less.

---

[1]These were computed from forecasts published in the *Federal Power Commission News*, Vol. 9, No. 29, July 16, 1976, p. 36. They had previously been supplied to the FPC as part of an April 1, 1976 response to Appendix A-1 of FPC Docket R-362, item No. 1.

Two remarks should be made about the 1969-1975 growth rates, however. First, the data incorporate only two years' experience with higher oil prices, and the structure of fuel consumption suggests that adjustments probably take a longer time. Second, this period saw unusually slow growth in GNP. The average annual growth rate in real GNP between 1969 and 1975 was less than 2 percent, while, for example, the average annual growth rate between 1963 and 1969 was almost 4.5 percent.[1] These two factors -- an expected continued adjustment to higher fuel prices, and a probable higher growth rate in real GNP -- work in opposite directions, so that it is not a simple matter to determine whether the 1975-1980 rate of growth in electricity consumption is likely to be faster or slower than the 1969-1975 growth rate.

For this reason, and because I felt that the forecasts made by the utilities themselves might be more accurate than historical growth rates during a period of sudden changes and considerable uncertainty, I used the forecasts of the reliability councils rather than forecasts based on historical growth rates.

Nuclear Generation

I forecast nuclear generation in 1980 by adding to 1975 nuclear generation an estimate of generation from plants that might be expected to be in operation by then. The assumption used in Chapter 4 was also

---

[1]Calculated from data in *Economic Report of the President*, January 1977, Table B-2, p. 188.

used here: only nuclear plants currently expected to be in operation by 1978 will be in operation by 1980, and they will be operated at a 70 percent load factor.[1]

Subtracting nuclear generation from net electricity generation yields an estimate of 1980 fossil net generation. The ratio of this figure to 1975 fossil net generation, multiplied by 1975 coal use, results in the extrapolated estimate of 1980 coal use. This procedure implies, in the states using Appalachian coal, an annual compound growth rate in utility demand of 4.6 percent between 1975 and 1980. Total Appalachian coal production, according to this forecast, will be about 494 million tons by 1980, roughly 20 percent higher than its 1970 level. Detailed figures on utility consumption by state and Appalachian coal production by district are presented in an appendix to this chapter.

Impact on Coal Consumption: Conversions from Oil to Coal

An alternative approach to the extrapolation described above is to estimate 1980 utility coal consumption as the sum of coal burned by existing plants, planned plants, and plants converted from oil to coal. Because there is considerable uncertainty about the extent of the reconversions from oil to coal, in this section I estimate that portion of coal consumption in several different ways. Because that component

---

[1]In fact, very few nuclear plants are scheduled to begin operation between 1978 and 1980, so that the assumption of a 2-year slippage is not materially different from the assumption that no slippage occurs. See National Coal Association, *Steam-Electric Plant Factors*, 1975 edition, Table 5.

is the only one that differs among the forecasts described in this section , I first discuss briefly the method used to estimate coal burned by existing and planned plants. I then present the range of estimates of coal burned at converted plants.

For the sake of simplicity, I assumed that plants already burning coal in 1975 would burn the same amount of coal in 1980. This assumption implies that no plants presently burning coal will convert to oil and that they will continue to be operated at the same capacity utilization rate over the next 5 years. Although this assumption may, therefore, tend to overstate coal consumption at these plants, due to a gradual reduction in the use of older plants as newer plants are brought into operation, the reduction over a 5-year period probably is not great enough to warrant an adjustment.

Coal consumption at plants planned to burn coal and intended to be put into operation between 1975 and 1980 was also estimated very simply. Because of the length of time required for planning and construction of a large coal-fired plant, it is virtually certain that only plants announced by 1975 can be in operation by 1980. I therefore assumed that all plants planned for operation by 1980 would be con- structed on schedule and that, as new large plants, they would be run as base-load plants.[1] Consequently, I assumed a load factor of 70 percent (the same load factor assumed for new nuclear plants). In estimating the coal consumed at these plants, I assumed a heat rate

---

[1]This assumption may overstate coal consumption, especially if the slackening of growth in electricity consumption leads to delays in con- struction or cancellations of announced plants. There is no simple way to adjust for these possibilities. The assumption thus represents an upper bound on coal consumption at new plants.

of 9000 BTU per kilowatt hour (a low heat rate characteristic of modern and efficient coal-burning plants). I further assumed that the heat content of the coal burned would be the same as the average heat content of coal burned by utilities in the same state during 1974 (the most recent year for which data were available).

Estimating coal that will be burned by plants converting from oil is more problematical. Because the physical equipment required to burn coal is much more extensive than that needed to burn oil, the decision to change an oil-burning plant into a coal-burning one is a complicated investment decision that depends on the remaining economic life of the plant, the conversion costs, and the additional costs of equipment needed for pollution control (principally sulfur emissions but also particulates). In general, if the plant was originally designed to burn oil, it is not economic to convert it to coal. Frequently there is no storage area for coal, and transportation facilities as well as coal handling, crushing, and ash disposal facilities would have to be constructed. Even if it were economical to undertake such a large-scale project, it is not clear that it could be completed by 1980.

Consequently, most plants considered for conversion to coal were originally constructed either to burn coal or to burn both coal and oil. Even with these plants, there will typically be some investments required to restore the coal handling equipment, in addition to the pollution control equipment.

210

The task of deciding which plants could feasibly be converted to coal has been made considerably simpler by the Federal Energy Administration. This agency, by virture of the Energy Supply and Environmental Coordination Act of 1974, has the authority to order conversions. In this connection, the agency published a list of plants that could potentially convert from oil to coal.[1] At the end of June, 1975, they issued Notices of Intention (NOI) to issue conversion orders (technically, "Prohibition Orders", because they prohibit the burning of oil or gas) affecting, in the states burning Appalachian coal, 44 units at 19 plants, or a total of 8014 megawatts of capacity.[2] This group of notices was only the first stage of the process, in two senses. First, a fairly lengthy series of steps (including public hearings, draft and final environmental impact statements, and evaluations by the FEA) must occur before the conversion orders take effect. Some of the proposed orders are being opposed by groups and individuals concerned about the impact of the conversions on air quality as well as by utilities expecting the orders to result in higher costs.[3] Second, the FEA expects to issue conversion orders for additional plants.[4]

---

[1]This list, divided into groups roughly according to the ease of conversion, was published in the *Congressional Record -- Senate*, May 5, 1975, p. S7419.

[2]The list of prohibition orders has been published in several places. This tabulation was taken from Office of Coal Utilization, Federal Energy Administration, *Implementing Coal Utilization Provisions of Energy Supply and Environmental Coordination Act*, April 1976, Appendix I-1.

[3]For example, at the public hearing on the draft environmental impact statement in connection with the prohibition order for the Schiller plant in New Hampshire, the Public Service Company of New Hampshire testified that costs would, on balance, increase if it had to convert to coal. I am indebted to Richard Meister of the FEA for discussion of this example.

[4]*Ibid.*, pp. 13-14.

There is thus considerable uncertainty about the number of conversions that will occur before 1980. The FEA list, however, provides an upper bound on the number of conversions, since it is probable that their estimate of the plants able to convert errs, if at all, on the liberal side. A lower bound is provided by assuming that no plants convert.[1] I also estimated coal burned at converted plants using the assumption that any plant able to convert (that is, any plant on the FEA list) would do so if the total costs of burning coal (including conversion and pollution control costs, as discussed below) were less than those of burning oil.

The Range of Estimates

Using this approach, therefore, I constructed 5 sets of estimates. The low estimate assumes that no conversions occur and that the utility coal burn in 1980 consists of the 1975 coal burn plus that burned by plants planned to burn coal. The high estimate is based on the assumption that all plants on the FEA list of those potentially able to convert do so by 1980. The three intermediate estimates reflect three different assumptions: (1) that all of the plants ordered to convert in June 1975 do so by 1980; (2) that plants able to convert

---

[1]In principle, this is not really a lower bound, since some plants may still convert from coal to oil, chiefly to comply with environmental standards. Such plants are potential candidates for conversion orders (*ibid.*, p.16). Given the relative prices of coal and fuel oil, however, switches from coal will probably occur only in isolated instances and hence not lead to much reduction in coal consumption. The assumption that no such conversions occur is probably a reasonable lower bound.

do so if it is economic to do so, and that total conversion costs
are at the low end of the current range of estimated costs; and (3)
that plants able to convert do so if it is economic, and that total
conversion costs are at the high end of current estimates.

In the rest of this section, I first discuss the available estimates
of conversion costs, their components and probable range, and then
I present the different estimates of utility coal burn and Appalachian
coal consumption in 1980.

Conversion Costs

Conversion costs include the costs of new equipment for flue gas
desulfurization (FGD), of upgrading the electrostatic precipitators if
necessary, and of restoring the facilities for handling coal. These
costs differ, therefore, from the costs associated with a newly designed
coal-burning plant -- retrofit costs for FGD are generally higher than
the costs of integrally designed equipment, but restoration costs of
coal handling equipment are less than new equipment costs.

These costs are subject to a great deal of uncertainty; as to
date there has been little experience with FGD or with reconverting
to coal. Flue gas desulfurization is especially problematical,
because the effectiveness as well as the costs have yet to be determined.
Costs also vary widely from plant to plant, depending on site conditions,
disposal facilities, age and condition of coal handling equipment,
and so forth.

In view of these uncertainties, I simply use the cost estimates
published by the FEA in their Final Environmental Impact Statement sub-

mitted in connection with the Coal Conversion Program under the Energy Supply and Environmental Coordination Act of 1974, Section 2.[1]

I first present the costs in general and then describe how these representative costs were translated into costs at the individual plants potentially able to convert to coal by 1980. Although for most of the costs I use a single estimate, the capital and operating costs of flue gas desulfurization are so large and so little known that I use both the low and the high estimates, leading to two estimates of conversion costs.

Costs in General

Conversion costs in general, divided into capital and operating costs, are shown in Table 7-6. Some of these figures merit discussion. The range of capital costs for FGD summarizes "estimates of FGD costs by the Environmental Protection Agency based on testimony in public hearings conducted on the status of the technology".[2] The witnesses included experts testifying on behalf of utilities, both public and private, and EPA consultants. The range spans several kinds of systems, and reflects engineering estimates as well as a few figures based on operating experience. It is thus both a reasonably recent and comprehensive survey of these costs, but it is by no means certain that the costs for all plants would lie within the range shown.

---

[1]Office of Fuel Utilization, Federal Energy Administration, April 1975, 2 volumes.

[2]*Ibid.*, p. IV-112. The hearings in question are Environmental Protection Agency, *National Public Hearings on Power Plant Compliance with Sulfur Oxide Air Pollution Regulations*, January 1974.

Table 7-6
ESTIMATED COSTS OF CONVERTING ELECTRIC
UTILITY PLANTS FROM OIL TO COAL, 1975

Capital Costs
($/kW)

Item	Low	High
Restoration of coal handling facilities	7.50	7.50
Upgrading of electrostatic precipitators	20.00	20.00
Flue gas desulfurization equipment	39.00	108.00
Solid waste disposal (FGD)	7.00	7.00
Total capital costs	73.50	142.50

Operating Costs
(Mills/kWhr)

Item	Low	High
Flue gas desulfurization	1.0	6.0
Excess of coal operating costs over oil operating costs	0.2	0.2
Total operating costs	1.2	6.2

SOURCE: Office of Fuel Utilization, Federal Energy Administration, *Final Environmental Statement, Coal Conversion Program, Energy Supply and Environmental Coordination Act of 1974, Section 2* (April 1975), pp. IV-109--IV-118.

215

The FEA in the source cited offers no rationale for the cost of upgrading electrostatic precipitators (ESP). Upgrading may be required because particulate removal is more difficult and important when coal is burned rather than oil, but also because use of low sulfur coal can reduce the efficiency of the ESP's.[1] Although these costs can vary widely from plant to plant, a great deal of research would be required to determine individual plant costs, so the FEA arbitrarily chose a single representative figure.[2]

The costs of restoring coal handling facilities are based on a January 1973 survey by the Federal Power Commission of about 725 fossil-fueled plants, adjusted for inflation.[3]

The operating costs for FGD are still controversial. EPA claims they run between 1 and 2 mills per kilowatt hour, while some utilities assert they are between 4 and 6 mills per kilowatt hour.[4] They include "the cost of auxiliary power, extraction steam for flue gas reheaters, limestone reagent for the sulfur dioxide absorbers (or equivalent materials in non-limestone FGD systems), maintenance and personnel to operate the systems".[5] I used two different figures -- a low figure

---

[1]*Ibid.*, p. IV-111.

[2]*Ibid.*, p. IV-112.

[3]*Ibid.*, p. IV-109. The FPC survey, Form 36, is reported in *Draft Report of the Oil Savings Task Force - Electric Utility Industry*, November 25, 1974.

[4]*Ibid.*, p. IV-117.

[5]*Ibid.*

of 1.0 mills per kWhr; and a high figure of 6.0 mills per kWhr, the
highest figures claimed by a utility. This range is admittedly wide.
The FEA assumed a figure of 1.9 mills per kilowatt-hour, based on its
own assumptions of operating costs, and one utility reported experience
of 2.7 mills. The technology is still largely untested, however, so
that use of this wide range is probably safer than assuming the truth
to lie somewhere in between.

Calculation of Costs at Individual Plants

To compare the costs shown in Table 7-6 with the difference in oil
and coal costs, it is necessary to put them all in the same units. For
this reason I translate the capital and operating costs into cents per
million BTU's. This change in units requires more assumptions for
capital costs than for operating costs.

Equivalent costs per million BTU of capital costs in dollars per
kilowatt can be expressed as follows:

$$C_i = \frac{K \cdot r_i \cdot 100000}{(8.760)\ (LF_i)\ (HR_i)} \tag{19}$$

where:

C_i = capital costs in cents per million BTU, at plant i;

K = capital costs, dollars per kilowatt (from Table 7-6);

r_i = annual capital charges (interest plus depreciation),
as a decimal fraction;

8.760 = thousands of hours per year;

LF_i = load factor at plant i in 1974 (as a decimal fraction); and

HR_i = heat rate at plant i in 1972, BTU's per kilowatt hour.

217

Most of equation (19) is a straightforward conversion of units. The load factor in 1974 was taken as most representative of plant utilization rates; it might be expected to decline gradually over the remaining plant life, and a more sophisticated procedure might use an exponential decay rate to calculate a present discounted value of the load factor. Given the imprecision of the cost estimates, however, and the relatively high utilization rates of even very old coal-fired plants (since they cannot be efficiently used to meet peak demand), I chose this simpler assumption.

Because heat rates tend to reflect plant design and do not vary much over the lifetime of a particular plant, the 1972 rate was chosen as a recent year and conveniently available.[1]

Choice of a cost of capital rate was more problematic. The interest rate was, somewhat arbitrarily, taken as 10 percent per year. Depreciation was, for the sake of simplicity, calculated on a straight-line basis. To figure annual depreciation charges, however, it is necessary to know the expected remaining economic life of the plant. Although old plants are rarely formally retired, the effective life is around 40 years. It is not, however, simple to determine the construction date of the potentially convertible units. The FPC publishes the initial construction date of each plant, but, typically, units are added at much later dates. To determine the construction date of units able to convert requires tracking the changes in plant capacity through *Steam-Electric Plant Construction Cost...* in every year after the plant first comes on-stream.

---

[1] Federal Power Commission, *Steam-Electric Plant Construction Cost and Annual Production Expenses,* Twenty-Fifth Annual Supplement -- 1972 (April 1974).

Such a procedure would be tedious and time-consuming and promises small reward.[1] Instead of using this procedure for all of the plants potentially able to convert, I used it only for those plants ordered to convert, and discovered that no unit on the list was more than 15 years old. I decided then to estimate the remaining economic life of each plant as:

$$L_i = max \begin{cases} 25\,(=40-15) \\ 40 - A_i \end{cases} \qquad (20)$$

where:

L_i = remaining economic life of plant i; and

A_i = age of oldest unit of plant i.

These assumptions, then, along with equation (19) were used to translate the capital costs into cents per million BTU. Translation of the operating costs, already on the basis of cents per kilowatt-hour, was much simpler. This conversion involved dividing the low and high operating cost estimates in Table 7-6 by the heat rates of the individual plants to obtain figures in cents per million BTU.

---

[1]For example, it is reasonable to assume that no unit older than 20 years would be required to convert. Assume an annual interest rate of 10 percent. With an effective life of 40 years, and with straight-line depreciation, the depreciation charges on a new plant would be 2.5 percent per year and on a 20 year-old plant would be 5 percent per year. The total cost of capital for a new plant would thus be 12.5 percent per year and for a 20 year-old plant would be 15 percent per year. The difference seems small, especially given the uncertainty of the cost estimates.

These calculations showed that, even starting from the same operating and capital costs, the costs at individual plants vary considerably. For the entire sample of plants on the FEA list potentially able to convert to coal, the costs ranged, using low FGD costs, from 27.0 to 69.1 cents per million BTU's; at high FGD costs, the range was 86.3 to 160.9 cents per million BTU's. The low costs do not seem plausible, but they do emphasize the range of uncertainty.

If the sample is restricted to plants ordered to convert as of June 30, 1975, the ranges are somewhat less, but still substantial (from 28.3 to 43.2 for the low costs, 89.9 to 121.4 for the high costs). In the next section, I present estimates of the amount of coal that would be burned if all those plants converted to coal for which the estimated conversion costs were less than the difference between the price of residual fuel oil and the price of coal.

Forecasts: Conversions from Oil to Coal

Forecasts of consumption of Appalachian coal are summarized in Table 7-7 under the heading of "Post-Embargo Forecasts". The lower-bound estimate, which assumes that no plants reconvert from oil to coal, implies total 1980 Appalachian coal consumption of about 487 million tons. The estimates of coal consumption if conversions occur range from a low of 504 million tons (on the assumption that all of the FEA Prohibition Orders of June 1975 are made effective) to a high of 534 million tons (on the assumption that all of the plants potentially

220

Table 7-7

COMPARISON OF PRE-EMBARGO AND POST-EMBARGO FORECASTS
OF 1980 CONSUMPTION OF APPALACHIAN COAL

(Millions of Tons)

Forecasts: Pre-Embargo	Utility Consumption	Total Consumption
(1) Extrapolation, based on pre-embargo electricity consumption growth rates	283.07	521.30
(2) No sulfur emission standard, low residual fuel oil prices and transport costs	233.45	471.68
(3) 0.7 percent sulfur-in-fuel standard, low oil prices and transport costs	141.31	379.53
(4) 2.0 percent sulfur-in-fuel standard, low oil prices and transport costs	314.30	530.54
Forecasts: Post-Embargo		
(5) No conversions from oil to coal	248.34	486.57
(6) Extrapolation, based on post-embargo electricity consumption growth rates	255.93	494.16
(7) FEA Prohibition Orders of June 1975 made effective by 1980	265.84	504.07
(8) Oil-to-coal conversions made on economic grounds, high conversion cost estimates	267.14	505.37
(9) Oil-to-coal conversions made on economic grounds, low conversion cost estimates	288.73	526.96
(10) All potential conversions listed by FEA occur by 1980	296.05	534.28

221

able to burn coal -- by the FEA's criteria -- do so by 1980).[1] The
great uncertainty about conversion costs, especially sulfur emission
control costs, results in a relatively wide gap between the two
forecasts of coal demand if conversions are made on economic grounds
rather than by fiat. The high conversion cost estimates imply coal
consumption very close to that implied by the FEA Prohibition Orders,
while the low conversion cost estimates imply that almost all of the
plants potentially able to convert would find it economical to do so.

---

[1]These forecasts are somewhat higher than those published as the
reference scenario, $13 per barrel imported oil, in Federal Energy
Administration, *National Energy Outlook*, February 1976. Table IV-8,
p. 176, shows utility coal consumption by census region for 1974 and 1985.
Interpolation by use of the discrete compound annual growth rates shown
there implies utility coal consumption in 1980 of 413 million tons for
the states covered in the present study. This figure, in turn, is
between 99 and 88 percent of the figures corresponding to the post-
embargo figures in Table 7-7 (the detailed figures are shown in
appendix tables 7A-1, 7A-3, 7A-5, 7A-7, 7A-9 and 7A-11).
     The forecasts of Appalachian coal production in the FEA volume
range between 94 and 85 percent of the figures shown as the post-embargo
forecasts in Table 7-7 (Table IV-33, p. 203).

Competition From Western Coal

As briefly discussed in Chapter 2, stringent sulfur emissions standards cause low sulfur Western coal to be an attractive alternative in some areas to high sulfur Appalachian or Midwestern coal with flue gas desulfurization (FGD). A careful analysis would involve estimating the delivered cost of Western coal at all of the locations presently using or planning to use Appalachian coal. While such an analysis is feasible, it is beyond the scope of this study.

In this section, instead, I present some rough calculations of the impact of Western coal on Appalachian coal production, with alternative FGD cost estimates. Delivered costs of Western coal at five locations in the market area of Appalachian coal are shown in Table 7-8. For the purposes of these calculations, I used these figures to estimate delivered costs of Western coal in each of the states burning Appalachian coal. (The assignments of costs to states are shown in Appendix Table 7B-1.)

By comparing these costs with the 1975 as-burned coal costs, state average, plus low and high FGD costs, I determined whether coal from existing sources (plus FGD) was less expensive than Western coal. If it was, I assumed that the extrapolated 1980 coal consumption (discussed earlier in Chapter 7) would be from current non-Western

Table 7-8

ESTIMATE OF DELIVERED COST OF WESTERN COAL
TO SELECTED DESTINATIONS WITHIN THE
MARKET AREA OF APPALACHIAN COAL

Destination	Cents/Million BTU
Hartford, CT	146
Concord, NH	162
Chicago, IL	107
Detroit, MI	120
Cincinnati, OH	114

SOURCES: Hartford and Concord -- Martin B. Zimmerman,
"The Potential for Coal Use in New England",
in *New England and the Energy Crisis*, p. 176.

Chicago, Detroit, Cincinnati -- estimated as
the sum of minemouth costs from Zimmerman,
p. 176, plus transportation charges from
Federal Energy Administration, *National
Energy Outlook*, 1976, p. F-71-3.

sources. If Western coal was cheaper, I made estimates using two
alternate assumptions. The first assumption is that all coal-burning
plants, both current and planned, can use and obtain Western coal.
This assumption is an upper bound on the vulnerability of Appalachian
coal. The second assumption is that only coal additional to that
burned in 1975 would be displaced by Western coal. This assumption,
in a rough sort of way, reflects potential difficulties in developing
Western coal on a large scale, transporting it to Eastern markets, and
burning it in plants designed for other kinds of coal.

This procedure is very crude, as it makes a number of assumptions
about the appropriateness of state averages, the delivered cost of
Western coal, the supply responsiveness of Western coal and of the rail
transport network, and downward inflexibility of Appalachian coal prices.
While the results should be used with considerable caution, therefore,
they still provide an estimate of the vulnerability to Western coal of
Appalachian coal when flue gas desulfurization is required.

The potential impact of competition from Western coal is shown in
Table 7-9. With the assumption of high flue gas desulfurization costs,
high-sulfur Eastern coal is more expensive than Western coal in all
states. With the assumption of low FGD costs, Western coal is still
the low cost alternative in 10 of the 24 states, representing a potential
decrease in utility consumption of Appalachian coal of about 75 percent
from the straight extrapolation scenario. If, on the other hand, it
is assumed that only coal additional to that burned in 1975 is open to
competition, the potential exposure is substantially reduced.

225

Table 7-9

1980 APPALACHIAN COAL SHIPMENTS IN
COMPETITION WITH WESTERN COAL

(Millions of Tons)

Assumptions	Utility Coal	Total Coal
(1) Low FGD costs, all coal-burning plants able to switch to Western coal	72.28	310.51
(2) High FGD costs, all coal-burning plants able to switch to Western coal	0.00	238.23
(3) Low FGD costs, only new coal-burning plants able to switch to Western coal	226.30	464.53
(4) High FGD costs, only new coal-burning plants able to switch to Western coal	207.15	445.38

In sum, although the calculations here are very rough, they indicate that consumption of Appalachian coal by utilities will be very sensitive to stack gas desulfurization costs, especially if Western coal is available on a relatively flat supply curve and if the transportation system does not prove a bottleneck.

Conclusions

Comparison of the pre-embargo and post-embargo forecasts in Table 7-7 leads to three main conclusions. First, the slackening of electricity consumption, due largely to higher electric rates associated with OPEC's price increases, would, by itself, cause consumption of Appalachian coal to decrease by about 5 percent from the level it would have reached without either sulfur emission controls or further conversions from coal to oil. That is, using similar methods to extrapolate coal demand to 1980 from pre- and post-embargo levels shows a falling-off of about 25 million tons, or 5 percent. To the extent that the post-embargo extrapolations do not measure the long-run adjustment to the change in relative prices, the decrease would be even greater.

Second, the OPEC price increases have in fact prevented Appalachian coal output from dropping sharply in response to strict sulfur emission controls. Even without sulfur emission standards, the relative prices of residual fuel oil and coal would (at 1969 relative prices) have caused Appalachian coal output to be somewhat less than the lowest post-embargo forecast (472 million tons compared with 487 million tons). An effective 0.7 percent sulfur-in-fuel standard would, however, have implied a substantial drop in Applachian coal consumption (to 380 million tons), in the absence of the 1973-1974 oil price increases. The forecast of 380 million tons was based on estimates of stack gas desulfurization costs that, after the fact, appear to have been too low, so that without the OPEC price increases the impact of sulfur emission standards on coal consumption might have been even more severe than implied by this forecast.

In short, then, the OPEC price increases are estimated to cause 1980 coal consumption to be between 15 and 63 million tons above what it would have been even without any sulfur emission standards, or between 3 and 13 percent higher. If the strict sulfur standards presently mandated by statute are in fact implemented, however, the OPEC price increases can be credited with increasing 1980 Appalachian coal output by between 107 and 155 million tons, or between 28 and 41 percent, above what it would have been at 1969 relative prices.

Third, Appalachian coal output is quite sensitive to the sulfur emission standard in effect. The actual standards vary, of course, from state to state (and even from city to city). If, for illustrative purposes, however, we compare a nationwide 0.7 percent sulfur-in-fuel standard with a 2.0 percent one, the figures in Table 7-7 imply that

228

Appalachian coal output would have been roughly the same with a 2.0 percent standard at 1969 relative prices as it will be with a 0.7 percent standard at 1975 relative prices. Thus, it could be said that the OPEC price increases rescued the Appalachian coal industry from the impact of a strict sulfur emissions standard.

Finally, Western coal presents a very serious competitive threat, depending on the costs of flue gas desulfurization. If large quantities of Western coal can be mined, shipped, and burned at the cost levels now estimated, it may displace substantial amounts of Appalachian coal. While the analysis of this possibility here is very rough, a comparison of the figures in Tables 7-7 and 7-9 shows that it is a possibility not to be lightly discounted.

Appendix 7A

This appendix contains the detail of the summary estimates
presented in Chapter 7. Tables 7A-1 and 7A-2 show the detail
by state and by Appalachian producing district for the forecasts
in which no plants burning oil convert to coal. Table 7A-1
shows utility coal consumption, while Table 7A-2 shows production
and employment. Tables 7A-3 and 7A-4 show the same information
for the scenario in which utility coal consumption is extrapolated
from its 1975 level. Tables 7A-5 and 7A-6 present the detail for
utility coal consumption estimates based on the assumption that only
conversions ordered by the FEA occur. Tables 7A-7 and 7A-8 show
the detail for the scenario in which conversions occur when they are
economical, on the assumption that high conversion costs obtain.
Tables 7A-9 and 7A-10 are for the same basic assumption, using low
conversion costs. Tables 7A-11 and 7A-12 are based on the assumption
that all of the possible conversions listed by the FEA occur.

Table 7A-1

ELECTRIC UTILITY COAL CONSUMPTION, 1980,
BY STATES BURNING APPALACHIAN COAL,
NO CONVERSIONS FROM OIL TO COAL

(Millions of Tons)

State, by Region	Coal Burned
Connecticut	-
Massachusetts	0.49
New Hampshire	0.99
New England Total	1.48
New Jersey	2.33
New York	9.51
Pennsylvania	42.49
Middle Atlantic Total	54.32
Illinois	41.69
Indiana	40.82
Michigan	25.20
Ohio	51.98
Wisconsin	15.52
East North Central Total	175.21
Delaware	1.90
District of Columbia	0.09
Florida	10.05
Georgia	22.92
Maryland	3.68
North Carolina	19.75
South Carolina	5.77
Virginia	4.07
West Virginia	32.36
South Atlantic Total	100.59
Alabama	25.46
Kentucky	32.45
Mississippi	3.68
Tennessee	24.29
East South Central Total	85.89
GRAND TOTAL	417.48

Table 7A-2

1980 PRODUCTION AND EMPLOYMENT IN THE APPALACHIAN COAL INDUSTRY,
BY DISTRICT

NO CONVERSIONS FROM OIL TO COAL

(Millions of Tons)

District	(1) Other	(2) Utility	(3) Total	(4) Proportion of 1970 Output	(5) Employment (Man-Years)
1	16.53	35.44	51.97	1.114	5627
2	31.87	10.74	42.61	1.076	5927
3 & 6	20.81	42.90	63.71	1.273	8895
4	12.33	54.80	67.13	1.206	7021
7	42.40	0.81	43.21	1.164	6007
8	105.90	82.91	188.81	1.172	24431
13	8.39	20.79	29.13	1.420	3190
TOTAL	238.23	248.34	486.57	1.185	61098

Table 7A-3

ELECTRIC UTILITY COAL CONSUMPTION, 1980, BY STATES BURNING APPALACHIAN COAL, EXTRAPOLATION FROM 1975 LEVEL

(Millions of Tons)

State, by Region	Coal Burned
Connecticut	-
Massachusetts	0.64
New Hampshire	1.27
New England Total	1.90
New Jersey	2.41
New York	6.46
Pennsylvania	45.15
Middle Atlantic Total	54.02
Illinois	46.38
Indiana	43.14
Michigan	25.69
Ohio	62.98
Wisconsin	17.39
East North Central Total	195.58
Delaware	1.28
District of Columbia	0.12
Florida	7.33
Georgia	18.47
Maryland	3.28
North Carolina	21.74
South Carolina	6.56
Virginia	4.34
West Virginia	36.66
South Atlantic Total	99.78
Alabama	18.56
Kentucky	35.77
Mississippi	2.10
Tennessee	21.73
East South Central Total	78.16
GRAND TOTAL	429.44

Table 7A-4

1980 PRODUCTION AND EMPLOYMENT IN THE APPALACHIAN COAL INDUSTRY,
BY DISTRICT
EXTRAPOLATION OF UTILITY DEMAND FROM 1975 LEVEL

(Millions of Tons)

District	(1) Other	(2) Utility	(3) Total	(4) Proportion of 1970 Output	(5) Employment (Man-Years)
1	16.53	35.71	52.24	1.120	5656
2	31.87	10.91	42.78	1.081	5950
3 & 6	20.81	45.70	66.51	1.329	9285
4	12.33	63.12	75.45	1.355	7891
7	42.40	0.84	43.24	1.165	6011
8	105.90	84.73	190.63	1.184	24667
13	8.39	14.94	23.33	1.137	2555
TOTAL	238.23	255.93	494.16	1.204	62015

234

Table 7A-5

ELECTRIC UTILITY COAL CONSUMPTION, 1980,
BY STATES BURNING APPALACHIAN COAL,
FEA-ORDERED CONVERSIONS

(Millions of Tons)

State, by Region	Coal Burned
Connecticut	-
Massachusetts	0.49
New Hampshire	1.22
New England Total	1.71
New Jersey	3.02
New York	11.30
Pennsylvania	42.49
Middle Atlantic Total	56.80
Illinois	41.76
Indiana	40.82
Michigan	26.01
Ohio	51.98
Wisconsin	15.69
East North Central Total	176.21
Delaware	2.72
District of Columbia	0.09
Florida	12.25
Georgia	23.72
Maryland	8.41
North Carolina	21.22
South Carolina	5.77
Virginia	4.07
West Virginia	23.36
South Atlantic Total	116.03
Alabama	25.52
Kentucky	32.45
Mississippi	3.68
Tennessee	24.29
East South Central Total	85.94
GRAND TOTAL	436.74

Table 7A-6

1980 PRODUCTION AND EMPLOYMENT IN THE APPALACHIAN COAL INDUSTRY,
BY DISTRICT
FEA-ORDERED CONVERSIONS

(Millions of Tons)

District	(1) Other	(2) Utility	(3) Total	(4) Proportion of 1970 Output	(5) Employment (Man-Years)
1	16.53	39.82	56.35	1.208	6101
2	31.87	11.04	42.91	1.084	5969
3 & 6	20.81	46.37	67.18	1.342	9379
4	12.33	55.32	67.65	1.215	7075
7	42.40	1.40	43.80	1.180	6089
8	105.90	90.94	196.84	1.222	25470
13	8.39	20.95	29.34	1.430	3213
TOTAL	238.23	265.84	504.07	1.228	63296

Table 7A-7

ELECTRIC UTILITY COAL CONSUMPTION, 1980,
BY STATES BURNING APPALACHIAN COAL,
CONVERSIONS FROM OIL TO COAL ESTIMATED ON ECONOMIC BASIS,
HIGH CONVERSION COST ESTIMATES

(Millions of Tons)

State, by Region	Coal Burned
Connecticut	4.39
Massachusetts	3.67
New Hampshire	0.99
New England Total	9.05
New Jersey	2.33
New York	9.51
Pennsylvania	43.93
Middle Atlantic Total	55.77
Illinois	41.69
Indiana	40.82
Michigan	25.20
Ohio	51.98
Wisconsin	15.62
East North Central Total	175.31
Delaware	1.90
District of Columbia	0.09
Florida	12.38
Georgia	22.92
Maryland	3.68
North Carolina	19.75
South Carolina	5.77
Virginia	10.83
West Virginia	32.36
South Atlantic Total	109.68
Alabama	25.55
Kentucky	32.45
Mississippi	3.68
Tennessee	24.29
East South Central Total	85.97
GRAND TOTAL	435.78

Table 7A-8

1980 PRODUCTION AND EMPLOYMENT IN THE APPALACHIAN COAL INDUSTRY, BY DISTRICT

CONVERSIONS FROM OIL TO COAL ESTIMATED ON ECONOMIC BASIS, HIGH CONVERSION COST ESTIMATES

(Millions of Tons)

District	(1) Other	(2) Utility	(3) Total	(4) Proportion of 1970 Output	(5) Employment (Man-Years)
1	16.53	43.31	59.84	1.283	6479
2	31.87	11.02	42.89	1.083	5966
3 & 6	20.81	43.41	64.22	1.283	8966
4	12.33	56.64	68.97	1.239	7213
7	42.40	1.46	43.86	1.182	6097
8	105.90	90.37	196.27	1.218	25396
13	8.39	20.93	29.32	1.429	3211
TOTAL	238.23	267.14	505.37	1.231	63328

Table 7A-9

ELECTRIC UTILITY COAL CONSUMPTION, 1980,
BY STATES BURNING APPALACHIAN COAL,
CONVERSIONS FROM OIL TO COAL ESTIMATED ON ECONOMIC BASIS,
LOW CONVERSION COST ESTIMATES

(Millions of Tons)

State, by Region	Coal Burned
Connecticut	4.39
Massachusetts	5.21
New Hampshire	1.27
New England Total	10.87
New Jersey	2.33
New York	21.24
Pennsylvania	43.93
Middle Atlantic Total	67.51
Illinois	43.07
Indiana	40.82
Michigan	25.89
Ohio	51.98
Wisconsin	15.62
East North Central Total	177.38
Delaware	3.44
District of Columbia	0.09
Florida	12.38
Georgia	23.78
Maryland	10.16
North Carolina	20.75
South Carolina	5.77
Virginia	10.86
West Virginia	32.36
South Atlantic Total	119.59
Alabama	25.55
Kentucky	32.45
Mississippi	3.68
Tennessee	24.29
East South Central Total	85.97
GRAND TOTAL	461.32

Table 7A-10

1980 PRODUCTION AND EMPLOYMENT IN THE APPALACHIAN COAL INDUSTRY, BY DISTRICT

CONVERSIONS FROM OIL TO COAL ESTIMATED ON ECONOMIC BASIS, LOW CONVERSION COST ESTIMATES

(Millions of Tons)

District	(1) Other	(2) Utility	(3) Total	(4) Proportion of 1970 Output	(5) Employment (Man-Years)
1	16.53	54.41	70.94	1.521	7681
2	31.87	12.95	44.82	1.132	6235
3 & 6	20.81	50.89	71.70	1.433	10010
4	12.33	55.46	67.79	1.218	7090
7	42.40	1.60	44.00	1.185	6117
8	105.90	92.44	198.34	1.231	25665
13	8.39	20.98	29.37	1.432	3216
TOTAL	238.23	288.73	526.96	1.283	66014

Table 7A-11

ELECTRIC UTILITY COAL CONSUMPTION, 1980,
BY STATES BURNING APPALACHIAN COAL,
ALL POSSIBLE LISTED CONVERSIONS FROM OIL TO COAL OCCUR

(Millions of Tons)

State, by Region	Coal Burned
Connecticut	4.39
Massachusetts	5.21
New Hampshire	1.27
New England Total	10.87
New Jersey	8.00
New York	21.24
Pennsylvania	43.93
Middle Atlantic Total	73.17
Illinois	43.07
Indiana	40.82
Michigan	25.89
Ohio	51.98
Wisconsin	15.62
East North Central Total	177.38
Delaware	3.44
District of Columbia	0.09
Florida	12.38
Georgia	23.78
Maryland	10.16
North Carolina	20.75
South Carolina	7.11
Virginia	10.87
West Virginia	32.36
South Atlantic Total	120.93
Alabama	25.55
Kentucky	32.45
Mississippi	3.68
Tennessee	24.29
East South Central Total	85.97
GRAND TOTAL	468.33

Table 7A-12

1980 PRODUCTION AND EMPLOYMENT IN THE APPALACHIAN COAL INDUSTRY,
BY DISTRICT
ALL POSSIBLE LISTED CONVERSIONS FROM OIL TO COAL OCCUR

(Millions of Tons)

	(1)	(2)	(3)	(4)	(5)
District	Other	Utility	Total	Proportion of 1970 Output	Employment (Man-Years)
1	16.53	55.90	72.43	1.553	7842
2	31.87	12.95	44.82	1.132	6234
3 & 6	20.81	54.71	75.52	1.509	10543
4	12.33	55.46	67.79	1.217	7090
7	42.40	1.87	44.27	1.193	6154
8	105.90	94.18	200.08	1.242	25890
13	8.39	20.98	29.37	1.432	3216
TOTAL	238.23	296.05	534.28	1.302	66969

242

Appendix 7B

This appendix presents the detail underlying the rough estimates
of the impact of low sulfur Western coal presented in the text
of Chapter 7.

Table 7B-1 shows the average as-burned cost of alternative
fuels in the states using Appalachian coal. The source of the
Western coal costs is described in the text; these prices are at
1975 levels. The base price of coal from existing sources in
these states is the average 1975 coal price. The low and high
estimates of flue gas desulfurization costs are 29.9 and 71.2 cents
per million BTU, respectively.

Tables 7B-2 through 7B-9 present the back-up detail on coal
consumption by electric utilities in the states burning Appalachian
coal and shipments from each of the Appalachian producing districts,
under alternative assumptions about the vulnerability of Eastern
coal markets and about the level of flue gas desulfurization costs.

ESTIMATED DELIVERED PRICES OF LOW SULFUR WESTERN COAL AND EASTERN COAL PLUS FLUE GAS DESULFURIZATION

(Cents per Million BTU's)

State, by Region	Western Coal	Eastern Coal & Low FGD Costs	Eastern Coal & High FGD Costs
Connecticut	146.0	160.8[1]	202.1[1]
Massachusetts	162.0[1]	160.8	202.1
New Hampshire	162.0	150.6	191.9
New England Total			
New Jersey	146.0[1]	188.7	230.0
New York	146.0[1]	147.6	188.9
Pennsylvania	114.0[1]	125.4	166.7
Middle Atlantic Total			
Illinois	107.0	105.3	146.6
Indiana	107.0[1]	89.1	130.4
Michigan	120.0	122.2	163.5
Ohio	114.0	125.1	166.4
Wisconsin	120.0[1]	116.3	157.6
East North Central Total			
Delaware	146.0[1]	145.2	186.5
District of Columbia	146.0[1]	179.4	220.7
Florida	146.0[1]	131.2	172.5
Georgia	146.0[1]	123.1	164.4
Maryland	146.0[1]	160.2	201.5
North Carolina	146.0[1]	137.3	178.6
South Carolina	146.0[1]	143.6	184.9
Virginia	146.0[1]	144.3	185.6
West Virginia	146.0[1]	117.1	158.4
South Atlantic Total			
Alabama	107.0[1]	121.6	162.9
Kentucky	107.0[1]	94.1	135.4
Mississippi	107.0[1]	112.0	153.3
Tennessee	107.0[1]	116.7	158.0
East South Central Total			

GRAND TOTAL

[1]Estimated from nearby state or state approximately same distance from coal fields.

Table 7B-2

ELECTRIC UTILITY COAL CONSUMPTION, 1980, BY STATES BURNING APPALACHIAN COAL, ALL COAL-BURNING PLANTS ABLE TO SWITCH TO WESTERN COAL, LOW FLUE GAS DESULFURIZATION COSTS

(Millions of Tons)

State, by Region	Coal Burned
Connecticut	-
Massachusetts	0.49
New Hampshire	0.99
New England Total	1.48
New Jersey	-
New York	-
Pennsylvania	-
Middle Atlantic Total	-
Illinois	34.09
Indiana	30.86
Michigan	-
Ohio	-
Wisconsin	11.65
East North Central Total	76.60
Delaware	0.97
District of Columbia	-
Florida	5.68
Georgia	14.62
Maryland	-
North Carolina	19.75
South Carolina	4.47
Virginia	4.07
West Virginia	26.23
South Atlantic Total	75.78
Alabama	-
Kentucky	25.59
Mississippi	-
Tennessee	-
East South Central Total	25.59
GRAND TOTAL	179.44

Table 7B-3

1980 PRODUCTION AND EMPLOYMENT IN THE APPALACHIAN COAL INDUSTRY,
BY DISTRICT
ALL COAL-BURNING PLANTS ABLE TO SWITCH TO WESTERN COAL,
LOW FLUE GAS DESULFURIZATION COSTS

(Millions of Tons)

District	(1) Other	(2) Utility	(3) Total	(4) Proportion of 1970 Output	(5) Employment (Man-Years)
1	16.53	4.17	20.70	0.444	2241
2	31.87	0.05	31.92	0.686	4440
3 & 6	20.81	14.18	34.99	0.699	4885
4	12.33	2.87	15.20	0.273	1590
7	42.40	0.55	42.95	1.157	5971
8	105.90	49.32	155.22	0.964	20085
13	8.39	1.14	9.53	0.464	1044
TOTAL	238.23	72.28	310.51	0.744	40256

Table 7B-4

ELECTRIC UTILITY COAL CONSUMPTION, 1980,
BY STATES BURNING APPALACHIAN COAL,
ALL COAL-BURNING PLANTS ABLE TO SWITCH TO WESTERN COAL,
HIGH FLUE GAS DESULFURIZATION COSTS

(Millions of Tons)

State, by Region	Coal Burned
Connecticut	-
Massachusetts	-
New Hampshire	-
New England Total	-
New Jersey	-
New York	-
Pennsylvania	-
Middle Atlantic Total	-
Illinois	-
Indiana	-
Michigan	-
Ohio	-
Wisconsin	-
East North Central Total	-
Delaware	-
District of Columbia	-
Florida	-
Georgia	-
Maryland	-
North Carolina	-
South Carolina	-
Virginia	-
West Virginia	-
South Atlantic Total	-
Alabama	-
Kentucky	-
Mississippi	-
Tennessee	-
East South Central Total	-
GRAND TOTAL	-

Table 7B-5

1980 PRODUCTION AND EMPLOYMENT IN THE APPALACHIAN COAL INDUSTRY,
BY DISTRICT

ALL COAL-BURNING PLANTS ABLE TO SWITCH TO WESTERN COAL,
HIGH FLUE GAS DESULFURIZATION COSTS

(Millions of Tons)

District	(1) Other	(2) Utility	(3) Total	(4) Proportion of 1970 Output	(5) Employment (Man-Years)
1	16.53	-	16.53	0.354	1790
2	31.87	-	31.87	0.685	4433
3 & 6	20.81	-	20.81	0.416	2905
4	12.33	-	12.33	0.221	1289
7	42.40	-	42.40	1.142	5894
8	105.90	-	105.90	0.657	13703
13	8.39	-	8.39	0.409	919
TOTAL	238.23	-	238.23	0.570	30933

Table 7B-6

ELECTRIC UTILITY COAL CONSUMPTION, 1980,
BY STATES BURNING APPALACHIAN COAL,
NEW COAL-BURNING PLANTS ONLY ABLE TO SWITCH TO WESTERN COAL,
LOW FLUE GAS DESULFURIZATION COSTS

(Millions of Tons)

State, by Region	Coal Burned
Connecticut	-
Massachusetts	0.64
New Hampshire	1.27
New England Total	1.90
New Jersey	2.33
New York	5.96
Pennsylvania	36.76
Middle Atlantic Total	45.05
Illinois	46.38
Indiana	43.14
Michigan	21.32
Ohio	46.87
Wisconsin	17.39
East North Central Total	175.10
Delaware	1.28
District of Columbia	0.09
Florida	7.33
Georgia	18.47
Maryland	3.28
North Carolina	21.74
South Carolina	6.56
Virginia	4.34
West Virginia	36.66
South Atlantic Total	99.75
Alabama	18.56
Kentucky	35.77
Mississippi	1.51
Tennessee	21.73
East South Central Total	77.57
GRAND TOTAL	399.36

Table 7B-7

1980 PRODUCTION AND EMPLOYMENT IN THE APPALACHIAN COAL INDUSTRY,
BY DISTRICT

NEW COAL-BURNING PLANTS ONLY ABLE TO SWITCH TO WESTERN COAL,
LOW FLUE GAS DESULFURIZATION COSTS

(Millions of Tons)

District	(1) Other	(2) Utility	(3) Total	(4) Proportion of 1970 Output	(5) Employment (Man-Years)
1	16.53	29.68	46.21	0.991	5003
2	31.87	8.96	40.83	0.878	5680
3 & 6	20.81	41.14	61.95	1.238	8649
4	12.33	49.33	61.66	1.107	6449
7	42.40	0.81	43.21	1.164	6007
8	105.90	81.81	187.71	1.165	24289
13	8.39	14.57	22.96	1.119	2514
TOTAL	238.23	226.30	464.53	1.112	58591

250

Table 7B-8

ELECTRIC UTILITY COAL CONSUMPTION, 1980,
BY STATES BURNING APPALACHIAN COAL,
NEW COAL-BURNING PLANTS ONLY ABLE TO SWITCH TO WESTERN COAL,
HIGH FLUE GAS DESULFURIZATION COSTS

(Millions of Tons)

State, by Region	Coal Burned
Connecticut	—
Massachusetts	0.49
New Hampshire	0.99
New England Total	1.48
New Jersey	2.33
New York	5.96
Pennsylvania	36.76
Middle Atlantic Total	45.05
Illinois	34.09
Indiana	30.86
Michigan	21.32
Ohio	46.87
Wisconsin	11.65
East North Central Total	144.79
Delaware	0.97
District of Columbia	0.09
Florida	5.68
Georgia	14.62
Maryland	3.28
North Carolina	19.75
South Carolina	4.47
Virginia	4.07
West Virginia	26.23
South Atlantic Total	79.55
Alabama	18.56
Kentucky	25.59
Mississippi	1.51
Tennessee	21.73
East South Central Total	70.87
GRAND TOTAL	341.73

Table 7B-9

1980 PRODUCTION AND EMPLOYMENT IN THE APPALACHIAN COAL INDUSTRY,
BY DISTRICT
NEW COAL-BURNING PLANTS ONLY ABLE TO SWITCH TO WESTERN COAL,
HIGH FLUE GAS DESULFURIZATION COSTS

(Millions of Tons)

District	(1) Other	(2) Utility	(3) Total	(4) Proportion of 1970 Output	(5) Employment (Man-Years)
1	16.53	29.32	45.85	0.983	4964
2	31.87	8.94	40.81	0.877	5677
3 & 6	20.81	35.65	56.46	1.128	7882
4	12.33	48.16	60.49	1.086	6326
7	42.40	0.77	43.17	1.163	6001
8	105.90	70.05	175.95	1.092	22767
13	8.39	14.26	22.65	1.104	2480
TOTAL	238.23	207.15	445.38	1.066	56097